- Market features and competition within the travel plaza/truckstop industry
- The role of technology in gaining industry leadership
- How international and national petroleum markets operate
- The major petroleum companies and the power of the industry
- Petroleum exploration, supply, and refining processes and products
- Insight into entrepreneurship and lessons on effective business management

THE FLYING J
STORY

From Cut-Rate Stations to the
Leader in Interstate Travel Plazas

An Authorized Biography and Company History

THE FLYING J STORY

From Cut-Rate Stations to the
Leader in Interstate Travel Plazas

AN AUTHORIZED BIOGRAPHY AND COMPANY HISTORY

BY

HOWARD M. CARLISLE

ABOUT THE AUTHOR

Howard M. Carlisle is an emeritus professor of business management, Utah State University. He served 20 years as department head and several years as director of the graduate MBA program. He is the author of seven books: two business management texts, a book for the American Management Associations, two family histories, and two biography/business histories. The first biography/business history, *Fulfilling a Dream: The Dee Smith Story,* (published in 1992) was on Smith's Food & Drug Centers, Inc. and its founder. The second, *Colonist Fathers, Corporate Sons,* the predecessor to this publication, was published in 1996. It contains "a selective history of the Call family" featuring Jay Call and the founding of Flying J. Inc.

ISBN 0-9651936-1-6

HMC Publishing
Logan, Utah

Printed in the United States of America
Cover design by Steve McRea

CONTENTS

FOREWORD

In early spring of 1968 when I incorporated Flying J for the purpose of building and operating live-in, self-service gasoline stations, I could not have predicted Flying J would evolve into the company it is today. I find it not only surprising but gratifying to provide the extensive services we offer to the public at large.

Flying J has always dealt in products that were easily available to the public from other sources. Our success has always depended on being able to deliver these products in a more efficient manner than our competitors. This has been possible only because of the people who became partners in this undertaking—whether as suppliers, short-term associates or those pursuing a career. It is our combined efforts, together with the wonderful opportunities made possible by the free enterprise system, that account for Flying J's success to date.

I think Howard has done a great job of telling our story and I hope you will find it interesting.

—Jay Call

PREFACE

In 1993 I reached an agreement with Jay Call to write a history of his company, Flying J Inc. I had authored several management books including a biographical business history, and I was looking forward to writing about another of Utah's most successful entrepreneurs of the past half century. As I started research on Jay's background, I discovered that he came from a long line of prominent Mormon colonists, many of whose descendants gained fame in the business world. I was fortunate to find that several colonists had written diaries and were included in short family histories that documented events in their lives. In addition, several of those who had established significant businesses were still alive and willing to sit through long interviews.

With this wealth of information, I received Jay's agreement to defer the Flying J book and write a Call family history covering his colonist ancestors, the business activities of his father and several other relatives, and the evolution of Flying J Inc. Three years later the book was published under the title of *Colonist Fathers, Corporate Sons*. The first two-thirds of the book is devoted to 150 years of Call family history followed by Jay's growing up years prior to beginning his business career. The latter third chronicles how he started Flying J and the various events leading to it becoming a company of national prominence.

In many respects, this deferral was a fortunate turn of events. From the start, current company president J Phillip (Phil) Adams wanted me to wait, predicting that the corporation was just at the point of attaining far greater achievements. In this respect, he could not have been more correct. Company sales in 1995 had reached $1 billion and now, six years later, they are $4.3 billion. In addition, the company has added a wide variety of services beyond fuel retailing that were unthinkable to most others before Flying J took the first steps. Although the business has grown to be the equivalent of a Fortune 500 corporation, Phil—anticipating greater things on the horizon—might still have liked me to wait a few years to see what heights the company would reach. However, to avoid what could be a never-ending process, he has concurred with Jay that publishing the manuscript at this time is appropriate.

As a professor, I came to be strongly convinced that understanding the basis for another person's actions or attitudes is limited unless one has a knowledge of the environment the person grew up in and the context of the situation in which these actions occurred. Individuals are influenced—to some degree molded— by other family members, their friends, schooling, local culture, economic conditions, and dozens of other factors. One can only comprehend why a person is angry, happy, despondent, or motivated by knowing the individual and being aware of the circumstances when these characteristics are displayed.

On the same basis, the progression of a company is best comprehended by having a knowledge of the business conditions that existed at each stage of its development. This requires information regarding consumer tastes, major competitors, how firms compete in the industry, technological trends, and a host of local, national, and—in the instance of petroleum—international economic and political data. As examples, actions of the OPEC governments to modify petroleum supplies will trickle down to influence what Flying J must charge for gasoline; a drop in national productivity and high unemployment will affect how much fuel consumers will purchase; high interest rates will force a company to put off planned capital improvements; if a competitor drops diesel prices by three cents, others selling this fuel are under pressure to respond; economies of scale make it almost impossible for small refineries to survive; and advances in telecommunications require companies to modify their operations and, in some instances, change their product lines.

Accordingly, this book is not a narrow history of an individual or an inside view of Flying J Inc. I have attempted to portray Jay's personal history and the development of the company in a broad context that considers the major environmental factors relevant to the time. To better acquaint the reader with one of the most significant of these factors, chapter 2 has been devoted to important happenings and trends in the economics of petroleum. In later chapters, considerable detail is given on Jay's background, his relationships with his father and the uncle who founded Maverik Country Stores, Inc., and factors that caused Jay to strike out on his own. At each stage of Flying J's development, I have sought to identify the external conditions that contributed to the ups and downs the company experienced. Based on this more holistic approach, the book provides insight into how the petroleum and transportation industries operate in relation to exploration, production, refining, retailing, marketing, trucking, and travel-plaza management.

An author is responsible for acknowledging individuals who have been of assistance in developing a manuscript. To avoid being tedious, I will not list the more than 100 family members, relatives, business associates, Flying J employees, and friends of Jay whom I interviewed or contacted in the search for information. Most of these names are included in various places throughout the book and can be found in the index.

I am especially indebted to Jay and officers of the company. They have given me open access to corporate records and responded when I was in need of more information. Following his management style, Jay has given me free rein to prepare the manuscript. Once he has confidence in someone in his employ, he prefers not to be bothered unless the person needs his assistance. His initial instructions were to write the book as an independent party, and the company has never taken any actions to prevent my doing this.

Special thanks go to Steve McRea who, again, designed a remarkable cover and to Tyler Leary who served as editor.

To keep the book more readable, I have used footnotes solely for informational purposes rather than to document findings. Those interested in documentation should feel free to contact me. My e-mail address is hmc@cc.usu.edu.

Howard M. Carlisle
Providence, Utah
February 2002

FLYING J: THE COMPANY TODAY

"Entrepreneurs who have built their own businesses . . . are continually on an economic roller-coaster personally experiencing the thrills and highs and devastating lows that are so much a part of the free-market system."

—Jon M. Huntsman,
Huntsman Corporation

One of the marvels of the modern world is how commerce flows around the globe. Goods produced in one area can be delivered locally within minutes and internationally within hours. If one could view all parts of the world from several thousand feet in space, the earth would appear as a hyperactive anthill. Steady streams of automobiles, trucks, airplanes, ships, and trains would be observed crisscrossing the earth's land and water masses in every direction. Fruit grown in Chile is purchased by consumers in Denver within 48 hours; lettuce sent from California's San Joaquin Valley is part of a salad for a family dinner in Salt Lake City within two to three days; automobile parts fabricated in

Detroit are shipped to a dealer the next day; and a business executive from St. Louis can be halfway around the world without losing a night's sleep. Transportation systems are to the market economy what the circulatory system is to the human body. Both constitute vital lifelines that make possible the product exchanges keeping each entity healthy.

Petroleum is the energy source that drives transportation. Petroleum provides 40 percent of the world's energy and 38 percent of that consumed in the U.S. The various modes of domestic transportation are just short of being totally dependent on petroleum products for their power with 95 percent of all mechanized transportation propelled by these fuels. Of the $270 billion spent in this country each year for petroleum, two-thirds is devoted to transportation.

American industry since World War II has been dominated by companies that supply petroleum and those that build vehicles (automobiles and trucks) powered by this precious commodity. Their long-run superiority is evidenced in the annual Fortune 500 list of the 500 largest U.S. corporations. Since *Fortune* first published this ranking in 1954, the top three companies each year have been petroleum or automotive corporations with the exception of U.S. Steel in 1954 and Wal-Mart in 1998 through 2000. Over these 47 years, petroleum companies have captured 41 percent of the top three positions. As recently as the early 1990s, petroleum remained the biggest business in the U.S. with more than 20 percent of the nation's stockholders' wealth invested in petroleum companies such as Exxon, Mobil, Texaco, Chevron, Shell Oil, and Atlantic Richfield.

Soon after World War II, the auto and petroleum industries turned into powerful oligopolies where a few firms shared each market and created barriers to entry that kept newcomers out. In the 1970s, Japanese and German automotive firms broke up the domestic oligopoly held by the big three automotive manufacturers, and the power of the major U.S.-based petroleum companies was diminished when domestic production became inadequate and imports became a necessity. Even though the power of these oligopolies was weakened, no domestic start-up company has since been able to break into the select circle of firms running each industry.

Fortunately for Flying J, one segment of the petroleum industry—truck transportation—was overlooked by the petroleum giants. One difficulty experienced by most oligopolies is that the members become more concerned about protecting their flanks than in improving customer service or exploring new markets, especially when the primary product (in this case crude oil) remains in high demand and is extremely profitable. The leading oil companies (Texaco, Chevron, Mobil, Exxon, etc.) pursued fuel retailing but more as a secondary interest. In the 1960s and 1970s, the trucking industry lacked the prominence that it does today, and no one showed much interest in serving what was viewed as an undesirable clientele.

In many respects, this rejection is difficult to understand except for the conservatism characteristic of an industry dominated by a few key players. Trucking now constitutes one of the country's largest industries with annual revenues exceeding $400 billion. Trucks haul an incomprehensible $5 trillion in freight each year, 80 percent of the nation's total. In fact, for three-quarters of the communities within the country, this form of transportation is the sole means of shipping and receiving cargo. The dominance of trucking is such that a familiar saying evolved: "If you bought it, a truck brought it."

Given these market conditions, how likely is the prospect that a young man from a rural community in Idaho, beginning in 1960 with one four-pump leased gasoline station, would become the national leader in truckstops and travel plazas and, in the process, secure a place among the prestigious petroleum firms? How likely is it that this budding entrepreneur in 42 years would create a petroleum-related private company with annual sales of more than $4 billion, making it the equivalent of one of the nation's largest 500 public companies in *Fortune's* annual listing? Ignoring the market-entry problems created by a powerful oligopoly, consider the following in calculating such odds:

- Nine out of ten start-up companies fail or no longer exist within ten years.
- The number of businesses in the U.S. exceeds 24 million, with only 500 on the *Fortune* list.
- Some 31,000 gasoline stations have separate owners and more than 1,500 companies operate truckstops.

For someone to overcome these odds would require an Horatio Alger–like story with the entrepreneur facing especially difficult circumstances trying to horn in on a mature industry controlled by an oligopoly where the primary product (oil) is little different in 2002 than it was in 1960. Although the primary product has experienced little change, petroleum exploration, production, refining, and retailing operations have all been modified, especially by technology. Technology has resulted in quantum leaps in efficiency and in a variety of new services, primarily spawned by electronics. As one analyst wrote in 1997, "Indeed, the oil industry is closer to a high-tech industry than the commodity business it once was." Here again, Flying J has been a pioneer, leading in everything from electronic point-of-sale devices to numerous applications involving state-of-the-art telecommunication systems.

Before we begin the story, take notice of these remarkable measures of the company's current status:

- Flying J leads the U.S. in on-highway diesel fuel sales.
- It ranks 46th on the *Forbes* list of the largest U.S. private corporations.
- In the *Hoover's* ratings of all private, public, and governmental business enterprises, Flying J is listed as 493rd.
- In the Fortune 500 breakdown by industry, if Flying J were a public rather than a private corporation, it would be ranked as 15th in petroleum refining, the industry classification that contains most brand-name oil companies.
- With 157 modern, full-service travel plazas connected in a continent-wide network, Flying J is the premiere hospitality provider for the traveling public. (Flying J plazas are

located in all continental states except Minnesota and six smaller states in New England. The company has three plazas in Canada.)

- Flying J has led in applying technology to all facets of its operations and is continually adding services (banking, insurance, telecommunications, etc.) beyond those traditionally offered by fuel-retailing companies.
- The company has become a one-stop center to meet the transportation and financial needs of carriers, drivers, and other industry participants.

In the process of its development, Flying J has led the way in upgrading the entire industry. The company set a standard that transformed the dingy, grubby, limited-service truckstops of the 1970s into large, modern, attractive plazas that meet a variety of customer needs. No longer are services slow, restaurants of questionable hygiene, attendants surly, facilities in need of repair, or washrooms without towels or showers. Drivers, once weary from lack of comfortable facilities to rest after long stints behind the wheel, need no longer deal with unpaved parking lots with limited space, poor lighting, and questionable security.

Led by a founder who demanded cleanliness and service, Flying J started a trend that has revolutionized how travelers are treated on the interstate highway system. Years ago, those in automobiles had reason to avoid obtaining service at a fuel stop designed for 18-wheelers. Today at Flying J plazas, two-thirds of all customers drive four-wheel vehicles. Here they find low fuel prices, clean, tiled restroom facilities, a wide choice of quality, reasonably priced food, and hundreds of convenience items at plazas where hospitality and friendly service prevail. Once the dredges of petroleum retailing, truckstops have been turned into modern highway venues unrivaled in service, price, and quality.

Flying J's ride to the top was anything but smooth. With its leaders unwilling to go public with corporate stock, the company has constantly struggled to gain adequate financing. Occasionally in the early years, meeting the payroll was a challenge. In an industry where profit margins are extremely small, a company's growth is inhibited without significant outside financial support. In addition, planning is especially difficult in this industry where normal market economic forces are frequently overridden by political events. Fortunately, at least for Flying J, operating in a market with extreme swings between highs and lows creates opportunities as well as threats.

The Flying J story has many dimensions. It is a story built around a fascinating sequence of events that reveal how a young man with one year of college created a world-class petroleum retailing company—an adventure with many useful insights into entrepreneurship, business management, and how petroleum companies operate.

ECONOMIC FUNDAMENTALS
OF THE
PETROLEUM INDUSTRY

"Oil is the world's biggest and most pervasive business, the greatest of the great industries that arose in the last decade of the nineteenth century."

—Daniel Yergin,
Pulitzer Prize-winning author of
The Prize: The Epic Quest for Oil, Money, and Power

*F*lying J, during its more than 40 years as a corporation, has adjusted to gasoline prices doubling in a few days, tripling within weeks, and at other times, descending in the opposite direction just as rapidly. Due to lack of fuel, the company once had to close stations and look for profits in other markets to sustain its growth. Oil markets are known for their extreme volatility. Being subjected to a variety of political and economic forces, managers in this industry are like military commanders—they must plan for an assortment of scenarios to deal effectively with events that are highly unpredictable. It is definitely not a business for the faint-hearted.

Petroleum, the world's most important source of energy, has the dubious distinction of being a relatively simple commodity to extract and process, and yet, due to its military and industrial significance, countries and corporations are constantly struggling to protect their holdings. Small countries, rich in oil but poor in other natural resources, wield significant international power whereas populous nations with huge industrial complexes can be brought to their knees when short of oil. To appreciate the rise of industrialization and warfare in the last century, one must be aware of the role oil has played. Similarly, comprehending Flying J's ascent toward leadership in the retail oil industry is dependent on an awareness of the basics of how petroleum markets operate internationally, nationally, and on the local level.

The only commodity that exceeds oil in importance is food. However, food comes in a variety of forms and is generally marketed regionally whereas petroleum is essentially the same on every continent and priced internationally. Because a barrel of oil in Saudi Arabia is nearly identical to one in Houston, Texas, those selling petroleum products are forced to compete primarily on price plus the cost of transportation. The paradox of oil being a simple yet indispensable product has numerous ramifications for the entrepreneur who dares to step into the fray.

The following section summarizes the key events, a highlight reel of sorts, affecting the American oil business. Readers looking for a more comprehensive history might refer to *The Prize* by Daniel Yergin.

The Historic Rise and Fall of American Oil Power

The first U.S. oil well was drilled in Pennsylvania in 1859. Oil, a latecomer on the world commodity scene, provided the energy that was soon to drive industrialization, enabling machine power to replace animal brawn. Initially, the significant demand for oil was to produce fuel for kerosene lamps. Then, in 1898, the gasoline engine caused demand for refined fuel to skyrocket. Railroads and navies became powered by oil rather than coal because of price, ease in transportation and storage, and lower noxious emissions. The victory of the Allies in World War I stemmed in part from abundant American oil. Eighty percent of the oil used by Allied forces came from the American continent.

By 1920 the U.S. was the world's oil king, supplying two-thirds of global needs. In the 1930s, regional competitors to the U.S. arose when oil was first drilled in Mexico and Venezuela. In 1938, oil was discovered in Saudi Arabia—the region that was to replace the U.S. as the world leader in petroleum production.

In 1940, Japan relied on the U.S. for 80 percent of its oil. One reason behind the Japanese attack on Pearl Harbor was to limit American influence in the Pacific region, thus making it easier for the Axis powers to gain access to oil in the Dutch East Indies and adjacent lands. In World War II, more than 50 percent of all supplies handled by quartermaster units were petroleum products. The U.S. provided 90 percent of the Allies' fuel. A key component of military strategy in both World Wars was to deprive the enemy of oil. Germany's and Japan's eventual defeats were hastened by lack of fuel for planes, ships, tanks, and armored vehicles. As one prize-winning author stated, "If there was a single resource that was shaping the military strategy of the Axis powers, it was oil. If there was a single resource that could defeat them, that too was oil." Near the end of the war, Stalin toasted the U.S. by declaring, "This is a war of engines and octanes. I drink to the American auto industry and the American oil industry."

After the war, the U.S. remained dominant in oil production, providing more than half of the world's crude. However, with domestic demand growing, by 1947 America had begun to import petroleum. In 1950, oil surpassed coal (the energy source that spurred industrialization during its first 150 years) as the primary energy source in America and the world. Oil became the "black gold" of international exchange.

To this point in the rise of the petroleum industry, corporations (primarily American) dominated worldwide exploration, drilling, and refining. Governments began taking a major role when Mexico nationalized its oil operations in 1938, followed by Iran in 1951. In addition, for the first time, some governments passed regulations requiring big oil companies to share their profits. Still, as Middle East oil became more abundant, the "Seven Sisters" oligopoly (Standard Oil of New Jersey, Socony-Vacuum, Standard Oil of California, Texaco, Gulf, Royal Dutch/Shell, and British Petroleum) held a corner on exploration, production, and refining.

In 1960, when Standard Oil of New Jersey (later Exxon, now Exxon Mobil) cut prices by 14 cents a gallon, oil ministers in those undeveloped countries, rich in oil deposits but poor in other resources, met to decide how to limit the damage and to gain more control over pricing. Out of this meeting, OPEC (Organization of Petroleum Exporting Countries) was formed. Saudi Arabia, Venezuela, Kuwait, Iraq, and Iran were later joined by the United Arab Emirates, Nigeria, Indonesia, Libya, Algeria, Qatar, Equador, and Gabon. The formation of OPEC set the stage for politicizing production and pricing by placing control in the hands of government officials rather than business executives.

During the 1960s, oil prices dropped 20 percent (an inflation adjusted 40 percent). As a result, worldwide industrialization surged, raising the standard of living in most countries. In 1970, the U.S. reached its all-time peak in oil production. Nonetheless, oil dominance was rapidly shifting from the Western to the Eastern Hemisphere. America provided 21 percent of the world's oil, the Middle East 30 percent. By 1973, 36 percent of the oil and oil products consumed in the U.S. came from foreign sources. American oil production remained higher than warranted by world prices due to the Eisenhower administration imposing import quotas beginning in 1959. Because of the quotas, U.S. producers unnecessarily drew down national petroleum reserves and, desiring to keep prices high, they resisted actions that would open competition with foreign operators.

On October 6, 1973, Egypt attacked Israel, initiating the fourth Arab-Israeli war. In less than four months after the outbreak, OPEC oil prices quadrupled, going from $2.90 to $11.65 a barrel. With oil being the world's biggest business and the largest single item in world trade, these price shock waves shattered confidence in financial markets. The U.S. was especially affected due to Arab nations cutting off oil supplies to America because of its ties with Israel. Retail gasoline prices jumped 40 percent, and fuel shortages resulted in long lines at stations nationwide. In the next two years, the American gross national product (GNP) dropped 6 percent while the Middle Eastern countries enjoyed an enormous transfer of wealth (growing more than 300 percent in 12 months) thanks to oil exports. In 1975, oil prices alone forced the U.S. into a negative trade balance (the value of imports exceeding exports). From April of 1974 until mid-1978, oil prices increased less than inflation. However, officials in Washington, not wanting to see national economic and military security threatened by an inadequate oil supply, adopted numerous programs (such as the 55-mile-per-hour speed limit) to conserve this precious commodity.

The second Middle East oil shock wave, much slower to develop, required 18 months (late 1979 to early 1981) for prices to triple. However, the dollars spent for oil were much greater because the initial crude price was $13 dollars a barrel compared to $3 in 1973, and the price shot to $34, more than ten times that of a decade earlier. The first price jump occurred when the Shah was disposed in Iran in 1979, and war appeared likely when Iranian students took 63 American hostages. A year later Iraq attacked Iran, dropping their joint oil production by two-thirds— a significant reduction in world supply since these two countries provided 13.7 percent of global needs. Fortunately, many countries had built shock absorbers into their economic systems to soften such a blow. Regardless, governments, businesses, and individuals alike immediately took to hoarding, causing an artificial shortage and again forcing high prices and long lines at supply sources. The combination of high oil prices and fuel shortages contributed to the world's first significant recession in nearly a decade. The U.S. saw interest rates spiral to 21.5 percent—

choking off business expansion, causing consumers to purchase less, and swelling unemployment.

In 1979, a national decline in oil demand heralded a decade-long descent. Higher petroleum prices and government-initiated conservation programs (including 55-mile-per-hour speed limits and fuel efficiency standards for cars) combined to drop demand. The payoff of such measures was a 32 percent decrease in oil usage. Automobile fuel economy led the way with an impressive 55 percent reduction in seven years.

The largely self-imposed oil shortage disappeared in 1981, and petroleum prices took a downward slide for the next four years as demand fell off. In 1982, the OPEC cartel—now supplying more than 50 percent of the world's crude—responded to lagging prices by limiting production in hopes of forcing rates up. However, rising crude production in non-OPEC countries plus the decline in world demand offset the quota decrease, and prices continued their downward trend with Saudi Arabia taking the brunt of the losses. From 1980 to 1984 the Saudis' export income plummeted by a devastating 83 percent. With the value of OPEC oil exports dropping more than 75 percent, many member countries cheated on quotas to offset tremendous reductions in national income.

In 1985, OPEC let prices float. Accordingly, rates fell in early 1986 as rapidly as they had risen in 1973, partially because Saudi Arabia opened its production spigots to punish those OPEC countries that had violated quotas. Within a few months, prices plummeted from $31.75 to $10 a barrel. With prices in a tailspin, producers and retailers scrambled to avoid being caught with large crude supplies. By reducing inventories, they further flooded the market, causing the inflation-adjusted barrel price to plunge lower than it had been in 1947. From 1986 to 1990, petroleum prices remained relatively stable largely due to tranquility among the major producing nations. Then, in August of 1990, Iraq invaded Kuwait, resulting in the Gulf War. Prices spiked to $40 a barrel, but the war soon ended and rates fell to former levels.

By the early 1990s, American oil production had dropped to

11 percent of the world's output. OPEC members remained the leaders at 40 percent. OPEC's dominance was reflected in reserves with a commanding 75 percent of the total versus a meager 3 percent for the U.S. The global oil market was relatively calm from 1991 through 1998. Prices remained stable at approximately $14 per barrel. This price was actually 31 percent lower than the 1947 price on an inflation-adjusted basis. Cheap oil was one of the major factors contributing to this era of unparalleled world prosperity, especially in the U.S.

Beginning in 1994, only half of the oil consumed in the U.S. came from domestic sources, making the country strongly dependent on foreign suppliers to keep the economy ticking. Oil deficits with other nations accounted for approximately half of the country's worrisome negative trade balance that was then at an all-time high. On the positive side, American industry was becoming less dependent on oil as manufacturing was giving way to electronics, technology, service, and other "soft" industries. Between 1973 and 1999, the energy required to produce an inflation-adjusted dollar of gross domestic product (GNP) dropped nearly 50 percent. Oil was now just 3 percent of GNP versus 9 percent in the late 1970s, causing many economists to conclude that the American economy was now buffered from being hog-tied by a lack of oil. Accordingly, the petroleum industry, showing its maturity, increasingly suffered from slow growth with national oil consumption expanding at less than 2 percent per year. Industry analysts predicted that companies tied solely to oil would end their long reign as the kingpins of American industry.

Oil's importance was not lost, though, as the typical forces causing peaks and valleys in oil prices once more shook economic markets in the fall of 1998. In December of that year, a glut in world supply caused prices to drop to $10.25 a barrel, a 12-year low. The decrease sent the oil majors scampering to merge, reduce employment, and cut capital spending. However, by the latter part of 1999, as demand continued to increase and supply dropped, prices more than doubled to $26 a barrel, placing oil again on the front burner of national interest and causing petroleum suppliers' profits to rebound to their former heights. Domestic demand increased a full percent in one year, partially due to Americans driving larger, gas-hungry vehicles. The drop in supply was largely attributed to the actions of OPEC ministers who, in March of 1999, agreed to reduce production by 2.1 million barrels a day or 2.7 percent. (This time OPEC was a godsend as much as a demon for many revenue-starved petroleum producers that had suffered through the largest oil-industry recession in decades.)

One year after reducing production, OPEC, recalling its folly of pushing world prices too high a quarter century earlier, increased output, contributing to declining prices. Prices peaked in the fall of 2000 at more than $30 a barrel before gradually slipping to the $20 a barrel range in the last quarter of 2001. The two-year seesawing of prices confirmed that the international economy had not torn itself loose from the clutches of the OPEC suppliers, especially Saudi Arabia and United Arab Emirates, the only countries with significant excess capacity to meet the growing global demand. With supply tight and U.S. refiners operating at near capacity, even minor changes in supply or demand continue to have the potential of triggering major price movements.

How Firms Compete in the Worldwide Petroleum Industry

As noted, marketing of oil is unique because it involves a commodity that is produced, refined, transported, and priced worldwide and is subject to political forces that cause rapid fluctuations in supply. As has been evident since 1972, oil and politics make a volatile mixture. No petroleum company was adequately prepared for the oil crises of 1973, 1979, or 1985. As one leading economist stated after reviewing oil prices in the last half century, "The rise and fall of oil prices was one of the most spectacular and puzzling events in economic history."

Crude producers have reacted to this uncertainty in various ways by: attempting to gain the size and capital (through mergers or other means) to ride out long periods of unprofitability; spreading risks geographically through investing in production facilities around the globe; diversifying into nonpetroleum investments; and maintaining favorable relations with political figures in oil-rich countries.

The National Market:
Competition in Exploration, Production, and Refining

Flying J's current focus is being competitive in continental exploration, production, and retail markets. Obviously, the company cannot escape the capriciousness of global politics, but its competitors are basically those on the American continent. Here companies pumping and refining oil compete primarily through exploration (finding new sources), price, and to some extent quality (chemical content). Price is influenced both by location and economies of scale. As noted, a barrel of oil in Texas is obviously more valuable to a Houston refiner than a similar barrel in Saudi Arabia. Because of the importance of location, all refiners and major retailers (including Flying J) place significant emphasis on the transportation arms of their businesses.

Large refiners have an enormous cost advantage resulting from economies of scale. A plant with a 100,000-barrel-a-day refining capacity can produce oil products (gasoline, diesel, aviation fuel, and petrochemicals) far cheaper than a company that refines 10,000 barrels per day. Small refineries cannot compete unless they are located in an area protected from competition. The result is that a few large firms located on the Gulf and West Coasts dominate the industry.

Even with an oligopoly in control, the capital demands of exploration are sufficiently great that one firm will not often have the resources or be willing to undertake the gambles associated with major new drillings. An example is the Alaskan pipeline under ownership of five corporations. Likewise, medium-sized companies form joint ventures to share drilling risks, especially on undeveloped properties.

Competition in Petroleum Retailing

To understand the rise and fall of companies in fuel retailing, it is necessary to appreciate how firms compete in the industry. Their strategies contain seven components, outlined below.

Fuel Prices

Price stands out above all other considerations. Unleaded gasoline is unleaded gasoline to most consumers, and prices on a dealer's signage often determine where a buyer will shop. Flying J has consistently priced under competitors, giving the company extremely high volumes at each location. The problem created by cutthroat pricing is that industry profit margins are made extremely low. Satisfactory after-tax returns of 1 to 2 percent are considered favorable, a goal many companies rarely obtain. With such small margins, retailers have little wiggle room, putting pressure on to reduce operating costs no matter how minuscule these might be.

Location

Modern on-the-go consumers are ever on the lookout to minimize the time involved in maintenance activities such as keeping fuel in their vehicles. Hence, price can be of less importance if one can fuel up close by. This demand for quick service resulted in the explosive growth of the small retail C-stores (convenience stores) in the 1970s and 1980s. Finding well-located sites where traffic volume is high (such as freeway off-ramps and corners at main intersections) has become more critical to retailers, especially as the supply of undeveloped locations along the interstates diminishes near heavily populated areas.

Higher Margin Products

C-stores are not only convenient for fueling but customers save time in buying a quart of milk, loaf of bread, soft drinks, and numerous other items. With gasoline and diesel generally priced at little more than breakeven, C-stores strive to sell higher margin products with percentage returns five to ten times higher. Increasingly, plaza operators and C-store owners are relying on the variety and quality of fast foods, drinks, ice cream products, grocery staples, lottery tickets, attached restaurants, and a host of driver services as their profit generators.

Facilities

Years ago, the visual attractiveness of a facility was secondary if fuel prices were low. The assumption was the customer's only concern was quick service at a low price. Recently, especially with large travel plazas, the variety of services and a facility's visual appeal affect sales coming from both truckers and the traveling public. To be successful, truckstops must attract drivers by including showers, comfortable lounges, ample parking, numerous telephones, e-mail and Internet access, full-service restaurants, and other amenities.

Advertising

Advertising is less essential with petroleum products since the basic product—fuel—remains essentially the same. In addition, in comparison with other industries, new styles are not introduced each year and technical breakthroughs do not result in current products becoming obsolete within one or two years. In addition, seasonal changes do not need to be brought to the attention of customers. Years ago when the quality of fuel was uncertain, brand names were thought to represent quality, and advertising themes such as Texaco's "You can trust the man with the star" were popular. However, in the 1970s when fuel specifications became standardized, customers became more inclined to let price dictate their decisions.

Minimizing Risks

A C-store owner is extremely vulnerable to losses when another more modern, fully-equipped C-store opens next door. An owner of five stores in a small district or metropolitan area will find that all five properties will suffer if a local recession slows sales. Hence, a sensible strategy is to disperse outlets over several regions to prevent such an occurrence. Another common problem for small companies is being underfinanced. Price wars regularly occur, and unless an owner has sufficiently deep pockets to endure such combat, well-financed opponents have the opportunity to drive them out.

Transaction Time

In recent years, fuel retailers have relied on advances in electronics to reduce transaction time. Waiting in long lines at fuel desks is sure to test the demeanor of many customers. Long-haul drivers especially want to get in and out in a hurry. In responding to their needs, Flying J now has paperless transactions and uses two high-volume hoses to rapidly put 18-wheeler drivers back on the road.

In the balance of this book, the reader will note that many of Flying J's successes and failures are directly tied to the events and analysis contained in this chapter. The above conditions create two somewhat unique situations managers in the industry must face: First, in a market where the margin between cost and price is especially thin, management must be extremely disciplined and run a tight ship. Mismanagement, particularly that involved in failing to focus constantly on the bottom line, can shortly doom a company. Second, with frequent supply gyrations affecting pricing, a company must be postured to react quickly to unforeseen changes and have the staying power to weather what may turn out to be long periods of financial uncertainty. The key is developing a capacity to ride through these peaks and valleys and still keep the company on course.

A Pioneer Heritage of Enterprising Risk Takers

"It is worthwhile for anyone to have behind him a few generations of honest, hard-working ancestry."

—John Phillips Marquand

*S*everal questions inevitably surface in the biography of a successful entrepreneur. Where did the individual get the interest and incentive to start such a career? What traits made this person able to succeed? Where did these traits come from? To a degree, these questions harken back to the "nature-nurture" controversy that has gone on for centuries among the public at large and by scientists studying human behavior. The "nurture" proponents argue that individual traits are largely determined by the person's environment, especially those relating to early home life, while "nature" advocates claim that inherited genes control our makeup. In relation to these issues, Jay's life and those of his ancestors provide an interesting and informative case study. Jay's heritage includes an unusually large number of leaders who gained acclaim as colonists and pioneers. In turn, several of their offspring became renowned business entrepreneurs. To gain insight into some of the major influences affecting Jay in his early years, a review of his heritage deserves consideration.

Jay's Pioneer Forefathers

The beginnings of the Call line in America came from English pilgrims who settled in Massachusetts in 1637, some 17 years after the Mayflower landed at Plymouth Rock. Little is known of their histories, although one Call ancestor fought in the battle of Bunker Hill during the Revolutionary War and another in the War of 1812.

One of the first family members to leave the East Coast and move west was Cyril Call, a veteran of the War of 1812. Five years after the war, he took up residence in northern Ohio. In 1831, he became one of the first converts to the Church of Jesus Christ of Latter-day Saints (commonly referred to as Mormons). Joseph Smith, president and prophet of the church, and several of his disciples formed this religious order in upstate New York in 1830. The next year they moved church headquarters to Kirtland, Ohio. Although Cyril was part of the mass Mormon migration from Ohio to Missouri, back to Illinois, and then to Utah, it was his son, Anson, who gained fame as a colonist and frontiersman.

Anson's pioneer feats all occurred within the context of the Mormon migration west, thus necessitating a review of how this religious movement developed. In forming the church, Joseph Smith claimed that God directed him to reestablish the true church of Jesus Christ upon the earth following the failure of Catholicism centuries earlier. Hence, Mormons do not consider themselves as Protestants, but as a restored faith of Jesus Christ. Besides accepting the Bible as the word of God, Joseph came forth with another set of scriptures he reportedly transcribed from golden plates called the Book of Mormon (a book primarily devoted to a history of the American Indians).

Because Mormon beliefs were considered extreme and heretical, members of the faith were frequently ridiculed and abused. Rumors of polygamy (a practice started earlier but not formally acknowledged by church officials until 1852) added to the popular conviction that Mormons were at odds with the rest of society and must be made to conform to traditional social norms. Joseph Smith was tarred and feathered on several occasions, arrested 35 times, and eventually killed in 1844 while jailed in Carthage, Illinois. However, it was more than their unorthodox beliefs that separated Mormons from their neighbors. The church's extremely aggressive missionary program in New England and Great Britain resulted in hundreds of converts who rapidly swelled the ranks of any community where Joseph Smith and his followers resided. When Mormon settlements began to overflow with these new arrivals, local non-Mormons (referred to by Mormons at the time as Gentiles) recognized that they would soon become a minority and could be overcome by force or through the ballot box. In addition, Mormon military strength posed a threat since the sect did not consist of unarmed pacifists who responded to force by turning the other cheek. Within their ranks, Mormons developed a trained militia to protect them from gun-wielding mobs and possible Indian uprisings.

Difficulty with their neighbors forced Joseph Smith and his followers to leave Kirtland, Ohio, and settle in the sparsely populated northern counties of Missouri in 1838. It did not take long until their growing numbers again resulted in the state militia and armed mobs combining to drive them out. Within a year, Governor Boggs issued an order that Mormons "must be exterminated or driven from the state." At the time, Missouri was on the extreme western frontier. Because Joseph Smith and his followers were unprepared for a massive migration to the unknowns of the West, they turned around and marched back to Illinois, hoping to find refuge there. In a swampy bend on the Illinois side of the Mississippi River, they started the city of Nauvoo. Within five years, 15,000 Saints resided in the community, making it one of the largest in the state.

Again Mormon fanatical beliefs and the specter of their political dominance aroused mob action, this time resulting in the June 27, 1844, martyrdom of Joseph Smith and his brother Hyrum at Carthage Jail, 25 miles from Nauvoo. After Joseph's death, conflict between the Mormons and their neighbors continued, causing the Illinois legislature to revoke the Nauvoo charter and the Illinois governor to conclude that Mormons "will never be able to live at peace with their neighbors in Illinois."

Prior to his death, Joseph had prophesied that Mormons would eventually migrate to the Rocky Mountains. (In the same prophesy he proclaimed that Anson Call would "build cities from one end of the country to the other.") Fearing the death of other church leaders and uncertain of their future, most of the saints, led by Brigham Young, crossed the Mississippi and headed back through Iowa in the wintry spring of 1846, remaining in the area around Winter Quarters (Florence, Nebraska) until the following spring.

Brigham Young, soon to become Joseph Smith's successor as church president, was intent on getting his followers to an unpopulated region of the West where they would not have to worry about militant neighbors and could form their own kingdom without encroaching on other white settlers. In addition, he preferred to settle outside the United States, and thus the Great Basin region of the West was ideal. Then part of Mexico, this 1,000-mile-square land mass (consisting mostly of a desert) was bounded by the Rocky Mountains to the east and the Sierra Nevada on the west. (The basin covers all of Nevada, most of Utah and small portions of Arizona, Wyoming, and California.) With such a forbidding landscape, especially for agriculture, those who had previously passed through found little reason to set roots.

In the spring of 1847, Young led the first wagon train along the Mormon Trail (a course roughly paralleling the Oregon Trail), arriving in the locale that was to be named Great Salt Lake City on July 24, 1847. He was a visionary empire builder whose grandiose plans startled even his counselors. In 1849, less than two years after arriving, he declared to his followers:

> We will extend our settlements to the east and
> west, to the north and to the south, and we will
> build towns and cities by the hundreds, and thou-
> sand of Saints will gather in from the nations of the
> earth. This will become the great highway of
> nations.

Before the year was out, 26 towns were established as out-
posts in this 1,000-mile-square largely desert region that Brigham
Young had designated as the State of Deseret. Within a few years,
communities in faraway places such as San Bernardino, California;
Carson City, Nevada; Lemhi, Idaho; and St. George, Utah, were
established.

Soon the Saints were to face their old nemesis—the United
States government. A result of the Mexican War and the Treaty
of Guadalupe Hidalgo that followed was that the Great Basin
was made a U.S. territory on February 2, 1848. Within a year,
the national Congress cut Brigham Young's empire in half by
forming Utah Territory. (Eliminated were the areas in California,
Idaho, Arizona, and New Mexico.) As a concession to the
Mormons, the federal government made Brigham Young territorial governor.

In seeking to fulfill his expansionist plans, President Young
relied upon one of his most important lieutenants, Anson Call.
Anson had been a close associate of Joseph Smith's, often acting
as a messenger and as one of his confidants. He had bailed Joseph
out of jail in Kirtland, Ohio, gained prominence as a member of
the Nauvoo Legion, and opened his home to the Mormon
prophet on many occasions. However, Anson did not fulfill a
major leadership role until Brigham Young took over the helm of
the church.

In the second year of the Mormon migration (1848),
President Young led 2,500 Saints to their new Zion. Anson was in
charge of the first 20 wagons. With the trail well blazed from the

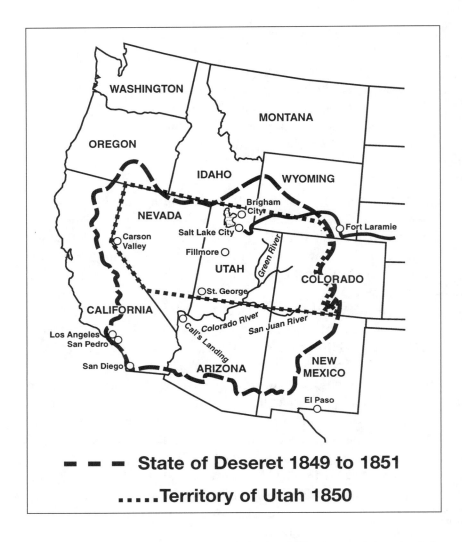

prior year, the company met with little difficulty, arriving in Salt
Lake City in the middle of September. Because of their late fall
arrival, Anson did not have time to build a permanent log cabin.
He traveled ten miles north of Salt Lake City to present-day
Bountiful and constructed an Indian wickiup, a cone-shaped hut
of willows and branches overlaid by a mixture of leaves, grass, and
mud. By the harvest of the second year, he had a log cabin for his
family and a yield of 1,000 bushels of grain.

Just as Anson was starting to enjoy the security of finally set-
tling down and ending the family's nomadic existence, Brigham
Young called him to help colonize an area 240 miles south of Salt
Lake City by establishing the city of Parowan, the southernmost
Mormon outpost at the time. Due to Anson's successful leadership as presiding elder in launching this community, Young then
assigned him to form a city that was to be the new capital of
Mormondom at a site 150 miles south of Salt Lake City. Under
Anson's direction, within three years, Fillmore, a community of
300 enterprising settlers, was established and the first wing of the
capitol building was in place. The building was never completed
to the original plans when, at this point, Young changed his mind
and kept the capital in Salt Lake City.

Anson's reputation as a colonizer led to more assignments
from the Mormon governor. Next, he spent the summer of 1856
in helping settle Carson Valley, Nevada, 600 miles west of Salt
Lake City at the foot of the Sierra Nevada. In the fall of 1864, he
undertook his most demanding assignment when Young
instructed him to lead a party to the Colorado River and establish
a port for ships coming upriver from the Pacific Coast. Young's
goal was to bring supplies and converts from the West Coast by
ship as far upriver as its depth and terrain would permit and then
convey them by land to settlements in Salt Lake City and other
Mormon communities. Anson selected a site in one of the most
formidable regions of the desert, 20 miles east of what was later to
become Las Vegas. The first boat docked at the port (Callville)
two years later, but the attempt to make river navigation possible
proved both costly and foolhardy. The port (now under the waters
of Lake Mead) was abandoned two years later, partially because it
was evident that the transcontinental railroad would become a
reality.

Anson, ever the opportunist, returned home to Bountiful,
Utah, and became engaged in the next undertaking that would
revolutionize life in the West. He obtained a contract to provide
a segment of the transcontinental railroad track in Weber Canyon
near Ogden, Utah. Following this, at the age of 60, he finally

settled down and concentrated on his own farming and business interests.

The Mormon colonization of the Great Basin was a remarkable undertaking in terms of its scope and planning, and Anson was one of the distinguished leaders in this monumental achievement. As the historian Juanita Brooks stated, Anson was "one of the great frontiersmen of Mormondom."

Anson had several traits that characterized at least two generations of his offspring. He was quick to adapt to new situations, pushed ahead regardless of what confronted him, and had hawk-like independence. Twice in Missouri he suffered severe beatings by mobs after ignoring Brigham Young's advice and attempting to gain compensation for his lost land and crops. He often acknowledged that he was "naturally stubborn and determined" and was a stern man of few words. Especially impressive was his ability to lead others and his "do it now" attitude.

A story that demonstrates his leadership and strong will comes from his first colonization experience when a party of 150 under George A. Smith, a Mormon apostle, was on the way to form southern Utah's first settlement. On a Sunday while stopped for religious services, apostle Smith noted that if only they had a bridge across the stream next to where they were camping it would be much easier for the many settlers expected to follow. To the apostle's surprise, when he awoke the next morning the bridge was in place. Anson had aroused several men at 1:00 a.m. after the Sabbath was over and had the wooden crossing installed within a few hours.

The settlement of the West is filled with stories of explorers noted for their rugged individualism, capacity to endure harsh climates, and being master of many trades. The unique trait of colonial leaders, such as Anson, is that they not only developed these survival skills but they got others to forsake this individualism and cooperatively participate in projects of mutual interest. More than a trailblazer, a colonist instills a sense of community where participants are willing to put their common interests first.

Anson and his first wife Mary Flint Call
(circa 1885)

Anson's Descendants: Anson Vasco I and Anson Vasco II

Let the reader be forewarned that names in the Call family tree are exceedingly confusing. Drawing pedigree charts of those in early Mormon history is often difficult because 10 to 15 percent of the males were polygamists, and those engaged in this practice typically had several wives and numerous children. Furthermore, the eldest children often were given the first name of a parent, thus adding to the confusion. In tracing Anson's descendants, we find he had 6 wives and 23 children. His eldest son, Anson Vasco Call, died at age 33 in Wyoming when returning from a Mormon mission in England. He left 2 wives and 10 children. His eldest son, Anson Vasco Call II (generally known as A.V.), had 4 wives and

47 children. To top off this confusion, A.V. named his first son Anson Vasco Call III, and the first and second names of most of his children came from other relatives.

A.V. was Jay's great-grandfather. When Anson Vasco I died, A.V. was left an orphan at age 12. By the time he was 33, he had three wives and five children. This was in 1883, one year after Congress passed the Edmunds Act that put teeth into the 1862 federal statute making plural marriage a criminal offense. Under the Edmunds Act, penalties were

Anson V. Call II

increased not only for polygamy but also for unlawful cohabitation. Federal marshals immediately filled territorial penitentiaries in Utah and surrounding territories with polygamists, thus forcing those with multiple wives into hiding.

Initially, to avoid arrest, polygamists crossed territorial lines into Idaho or Arizona, but federal marshals in these regions also stepped up enforcement, causing many violators to leave the country and seek refuge in Canada or Mexico. One of the local areas for polygamists to find a safe haven was in Wyoming. The governor there was actively seeking parties who would settle the region below Yellowstone National Park along the Idaho border and welcomed what other states considered "outlaw Mormons."

Thus, A.V., accompanied by one of his brothers and other family members, in 1887 became one of the first homesteaders of Star Valley, Wyoming, an unsettled 45-mile long, 10- to 15-mile-wide pocket high in the Rocky Mountains, 75 miles east of Fort Hall in Idaho. As a refuge, Star Valley provided polygamists a remote, difficult area to access. On the other hand, it was an extremely harsh environment to homestead. Local Indian tradition held that no one could live through the winter in this high

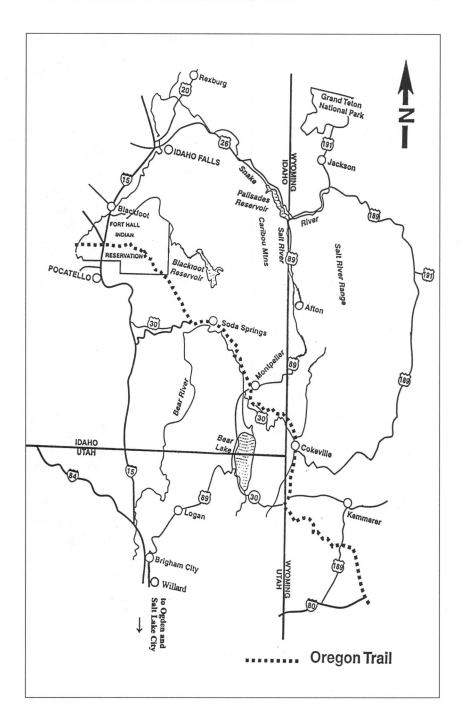

mountain terrain where temperatures often remained below zero and occasionally dropped to a minus 40 degrees. With the nearest railroad line 50 miles away and no major markets within 200 miles, supplies were difficult to obtain and farm commodities hard to exchange with outsiders.

A testament to the region's remoteness was that several outlaws wintered there between 1890 and 1894 with little fear of being captured. Well-known desperados such as Butch Cassidy, Tom McCarty, and Matt Warner each spent at least two winters in the valley. McCarty and Warner, posing as Montana ranchers, bought land and opened the area's first saloon. Their outlaw exploits are difficult to document, but they were reputed to have made several bank robberies late in the fall in locations such as Denver and Telluride, Colorado. After a holdup, they would return to Star Valley and hide out just prior to winter snows blocking the passes, making it difficult for posses to track them down. McCarty and Warner were on friendly terms with the locals. No one in Star Valley had enough money to make robbery worthwhile, and the outlaws openly fraternized with the settlers, eventually marrying two of the local maidens.

Anyone raised on the frontier was accustomed to hard work, lack of tools, difficult travel, severe weather, and inadequate food and housing, but A.V. and his families were still unprepared for the first years in their new homeland. Several settlers had attempted to become permanent residents beginning in 1879, but they gave up primarily due to the weather. The valley's first grain crop was not recorded until 1887, the fall that A.V. took his two oldest boys (one ten and the other eight) to build his first cabin and break up the sod, hoping to raise enough produce for his progeny to survive through the winter.

As an example of A.V.'s industry and his skill as a homesteader, during the first spring and summer, besides clearing land and planting crops, he built a cabin for each of his three wives and their ten children. He was constantly adding rooms and building new abodes for his rapidly expanding family. He became an extremely skilled carpenter and the valley's prominent builder.

Star Valley Stake House at completion in 1909.
Designed and constructed by Anson Call II

Within the next 30 years, he contracted to construct most of the schools, churches, and major homes in the area. In addition, he taught school, opened a hardware and furniture store, and served nine terms as mayor of Afton, the valley's major city.

A.V. shared many of Anson's traits: he was quick to act, proud, stubborn, and rarely backed down from a challenge. He taught himself complex engineering skills, and viewed hard work as the only way to get ahead. He was a strong disciplinarian, forcing his children to stand on their own at an early age. He became a master builder who saw that each of his sons become skilled in at least one of the associated trades. A common saying by valley residents was that "the Calls all have sawdust in their blood."

Grandfather Thomas Call

Jay's grandfather, Thomas Call, was a son of A.V.'s third wife. He became a builder, residing in Star Valley until after age 60. He constructed dozens of homes and commercial buildings at a time when slow growth in the area limited opportunities for local builders, many of whom were Calls. By 1950, the valley's population was just 5,000 with one-third living in the principal city of Afton. The average farm was a mere 200 to 250 acres, and industry was essentially nonexistent other than limited lumber, dairying, and cheese making. With a growing season of 50 days, only 15 percent of the land in 1950 produced grain crops. The balance was in hay meadows and pasture.

Star Valley's Spartan conditions resulted in the survivors being a sturdy stock of progeny who were hard working, frugal, and accustomed to adversity. Thomas worked long hours to provide a meager living for his wife and ten children. He sometimes went door-to-door seeking work as a carpenter and made money in the evenings by playing in an orchestra for $2 a night. Among his descendants, he gained fame for his honesty and frugality, traits that came to characterize his children and many of their offspring.

Tales of Thomas's frugality have been carried down through several generations. "Waste not, want not" was a common early Mormon theme, but Thomas carried it to an extreme, borne in part from his trying to find work in the Great Depression of the 1930s when U.S. unemployment soared to 25 percent. When working away from home, he slept in unfinished basements or in the construction storage sheds. A person was to always own, not rent, and workers were to be hired only if there was no other way the work could be accomplished. Every reusable material was to be salvaged. He paid his grandchildren 50 cents a day to collect old nails, straighten them out, and sort them by size. Boards were reused by cutting off the nail-hole portions at both ends. He was quick to tell others if he thought they were building a home too large for their needs and to chastise his daughters and daughters-in-law if they bought bread rather than made it. He even rejected his children's urges to get medical care in Salt Lake City because "it was too expensive." This frugality continued through several family generations.

The Early Entrepreneurs

Thomas shared two of A.V.'s traits that led to difficulties with his children. He could be stern, almost to the point of being abrasive. He was demanding of his eldest children, and a strict disciplinarian who would take his offspring to the woodshed when deemed necessary. Thus, at an early age, two of his oldest boys, Reuel and Osborne (Jay's father), were looking for ways to get out on their own.

When in high school, Reuel resolved to avoid the family tradition of being a builder and carpenter. Yet opportunities in Star Valley were largely restricted to the traditional ones of ranching and agriculture. Before graduating from high school, he demonstrated his entrepreneurial flair by buying into the latest technological fad. He purchased several dozen roller skates and rented large halls where he charged customers 25 cents an hour for the skates. Within two years he sold his business for $4,500 (approximately $45,000 in today's currency) and was on his way to fulfilling his entrepreneurial ambitions.

Reuel's next move was to take advantage of another new technology sweeping the nation. The automobile was just being introduced into the region, and these motorized contraptions relied on gasoline. After working one summer in a service station, Reuel decided to use his $4,500 to build a station at one of Afton's main intersections, the first of his many enterprises. Although he avoided being a carpenter, Reuel never lost the urge to build. Before long, he had several stations and soon became a wholesale distributor for Sinclair Oil, capturing a region from Yellowstone Park along both sides of the Wyoming-Idaho border to the Four Corners area of Utah, Colorado, New Mexico, and Arizona.

Within a few years, Reuel's string of stations was spread through a seven-state area. In addition, he was an owner or part owner of three refineries, and was one of the largest fuel wholesalers and truckers in the region. In 1968, he attained his milestone of selling one million gallons, followed five years later by a

Afton, Wyoming (Washington and Fourth Avenue), Christmas of 1938.
Reuel's Sinclair station in lower left.

peak of 231 million gallons, large in volume but marginal in profits. Soon after, his three sons took charge of the company. Within
a few years, they de-emphasized wholesaling to improve their balance sheet and concentrated on the retail chain of Maverik
Country Stores, a business now owned by his children and grandchildren that generates annual sales of more than $500 million
through 167 outlets.

Perhaps Reuel's most outlandish business scheme for this
sparsely populated region was to build a high-altitude airplane
suitable for flying locally. In 1932, at age 24, Reuel had earned his
pilot's license and, in a span of ten years, owned or was part owner
of ten airplanes. All were designed to take off and land at sea level,
making them difficult to navigate and land in the thin air of the
Rocky Mountains. Never one to pass up a business opportunity,
Reuel in his unconventional manner began contemplating ways of
solving this problem. Such a thought others would consider as

preposterous, even the machinations of a deranged mind, since aeronautical engineers, metallurgists, manufacturing engineers, machine shops, or support services of any type were located hundreds of miles away. Furthermore, Salt Lake City (200 miles away) was the nearest metropolitan area, and airplanes were not manufactured there. However, Reuel, ever oblivious to obstacles that others viewed as insurmountable, decided to push ahead.

Following an approach that became typical in his business endeavors, Reuel decided to involve his relatives. He contacted his uncle, Ivan Call, who was currently teaching physics at the local high school. Ivan had a master's degree in civil engineering but no background in aeronautics. When Reuel asked him if he could design an airplane, Ivan was obviously taken aback by such a request. However, Reuel's relatives had witnessed his accomplishments in the petroleum industry, and Ivan was not ready to write his nephew off as eccentric. After Ivan smiled and said, "Why not?" Reuel told him he had already ordered parts and was ready to develop a prototype through his new company—Call Aircraft.

Within a few weeks, a prototype was ready with Reuel in the cockpit as the proud test pilot. However, the Japanese attack on Pearl Harbor and the U.S. entrance into World War II forced plans to be delayed.

After the war, market conditions for introducing a new commercial plane were far more difficult. Manufacturers of military aircraft were converting to civilian production, and broad commercial development was underway. Reuel remained undeterred, and the company undertook design and production of four models in the newly named CallAir line. The standard two- or three-passenger model is a low-wing, cabin monoplane with a cruising speed of 110 miles per hour and a range of 650 miles. What distinguished the craft was its high horsepower-to-weight ratio, broad wings, and capacity to make sharp turns.

In the next 15 years before Reuel sold the company, Call Aircraft produced 12 to 52 airplanes annually. Demand was never adequate to keep the company in the black financially, but the venture always delighted Reuel, who often undertook projects of

interest with almost total disregard for the financial conse-
quences. Reuel eventually sold the company in 1960. Ownership
has changed several times since, but light utility aircraft and aero-
batic or stunt planes are still being produced in Afton. In total
more than 2,000 aircraft have been built in the area.

Reuel's genius as a pragmatic inventor is represented in other
innovations. Attachable skis were added to the CallAir, making it
possible to take off and land on snow without removing the
wheels, and a propeller-driven snow car, a forerunner of the snow-
mobile, was put into production. Though most of Reuel's inven-
tions were ingenious, he rarely made money on ventures outside
of the petroleum industry. Generally he was ahead of his time, a
difficulty that later characterized some of Jay's ventures. While in
his 80s, Reuel was still flying solo to his 21 C-stores in Utah,
Arizona, Nevada, and Wyoming. In 2001, he turned 93 and
remains active in business affairs.

In his early business ventures, Reuel was quick to get his
brothers and other relatives involved. He set one brother up as a
Sinclair distributor in Rexburg, Idaho, and he had Osborne (Jay's
father) build and then lease a station in Soda Springs, Idaho.
Osborne expanded this business by adding a bulk plant and
becoming a Sinclair distributor in this small community where Jay
was born and raised.

Osborne, born in 1917, was nine years younger than Reuel.
They were not only brothers but also the best of friends and eventually business partners. By the time Osborne graduated from
high school, Reuel was already one of Star Valley's most successful
businessmen. Osborne worked on several of Reuel's projects
including building struts for the first CallAir planes, and at Reuel's
request, he fabricated three small service stations in surrounding
communities. Osborne's drive and similar interest in converting
any reasonable opportunity into a business venture won Reuel's
respect and desire to keep his brother connected. Reuel served as
a significant, perhaps indispensable, role model, not only for
Osborne but also later for Jay. Jay became Reuel's favorite

Osborne (light shirt) in front of the Alpine, Wyoming, station he built for Reuel.
(July 1939)

nephew, many feeling that Reuel was closer to Jay than his own sons.

Many of the traditional Call traits—risk taking, frugality, strong will, fierce independence—served Reuel well. With these traits as a base, he developed into the new breed of entrepreneur that developed in the aftermath of World War II. As one admirer said, "He has an idea a minute," he is strictly a doer, not a bean counter, and is completely enamored with his work and lifestyle. He limits his focus to what interests him, is not guided by fixed rules, resents restraints on his freedom, and is driven to undertake new projects for the satisfaction of seeing them come into being rather than for financial reward. He is not one to flaunt his wealth or position. As one son stated, "He never spent a lot on himself except for airplanes." In this respect, he provides an important lesson for those going into business—successful, long-sustained entrepreneurship rarely is driven by greed.

Other Call Family Entrepreneurs

Reuel had several Call family contemporaries who also became successful entrepreneurs. However, they had no working relationship with Jay, making their ties entirely bloodline. Reuel and Osborne's father, Thomas, married Ethel Papworth. Her mother was Grace Christy Covey. The Covey family started some of the most successful early businesses in Utah and Wyoming. In 1886, Grace's grandfather formed a company with his sons that eventually had a herd of 60,000 sheep. When the federal government started restricting rangeland usage several decades later, the family divested their sheep operations and moved into other business interests. The resulting company built numerous apartments and business buildings in Salt Lake City. At one time, they had 364 apartments, making the Coveys the largest landlords in the area. In 1933, family members organized the Covey Gas and Oil Company and became Reuel's competitor. The company expanded into refineries and opened several prominent Little America motor lodges and hotels.

Another notable Call entrepreneur was Christian Joseph Call, one of A.V.'s 37 children. At an early age, Christian bought A.V.'s hardware store in Afton, operating it until 1926. In the next few years, he engaged in a variety of enterprises in Idaho and Utah, primarily in the grocery business. At age 55, he sold his holdings in the area, moved to California, and invested in the Sav-On Drug Company. After becoming president, he expanded the business into a chain of 100 stores, all but one in California. Sav-On was one of the first drug companies to offer a broad line of goods in addition to drugs and notions. The chain became one of the largest in California, all the more remarkable because Christian pushed the company to those heights in just over a decade before dying January 11, 1958.

GROWING UP IN A RURAL COMMUNITY AND STARTING A BUSINESS CAREER

"It is a wise father who knows his own child."
—Shakespeare

*L*ike Thomas's other children, Osborne learned the building trades from his father before graduating from high school. One year after graduation Osborne married a classmate, Janice Miller, who had just finished high school. A daughter, Sharon, was born to the couple one year later. On occasion while Osborne was working on odd jobs such as wallpapering, Janice would assist him while the newborn slept in a basket close by. Jay was born in 1940 at the time Osborne was constructing the Soda Springs station. Soon after, Osborne built a four-room house for his family several blocks away. The bathroom consisted of a basin with a cold tap and toilet. Bathing took place in the kitchen using a portable galvanized tub filled with water heated on a metal stove.

Like Reuel, Osborne devoted every spare minute to his new

Osborne (holding Jay), Janice, and daughter Sharon. August 1940.

business. With no working capital, his only choice was to make the station generate sufficient cash flow to cover mortgage payments. Soda Springs offered several commercial advantages over Afton: it was lower in altitude, making the growing season longer; the Oregon Trail had passed through the community and Soda Springs remained a major traffic route to the Northwest; and the area had railroad service. However, like Afton, it offered few attractions for someone intending to start a business. The area was sparsely settled with low-income farmers. The largest city, Soda Springs, could only boast of 2,000 residents. Furthermore, the nation was not yet out of the Great Depression, and impoverished farmers sought to obtain fuel on credit rather than pay cash. This situation especially affected Osborne since, as close as the Call brothers were, Osborne could not look to Reuel for financial help. Family members never expected special favors from other relatives. In fact, they would often set up business in each other's territory. Once when Osborne was having difficulty paying his fuel bill,

Reuel left him a note stating, "If you're out of gas and out of money, you're out of business." This Call legacy of expecting family members to stand on their own later carried over to Osborne's relationship with Jay.

Fortunately Osborne was a skilled deal-maker. He had the uncanny ability to envision and negotiate barter arrangements with suppliers and property owners, thus eliminating his need to put cash up front. As soon as he broke ground for the station, Osborne was exchanging construction materials and services for the future delivery of gas. Using this approach, he not only cut down on required capital but locked in customers. Later, when he owned an automobile agency, five years in a row he swapped an associate a new car in exchange for a local building. When he became owner of a motel, he arranged a similar fuel barter agreement with the company providing laundry service. And, as was quite often his practice, he purchased a choice site for a station with a new pickup as a down payment.

It was not just Osborne's skill as a deal maker that quick-started his business. He displayed unusual energy in acquiring and retaining customers. Once when a lumber truck drove through town, he jumped into his pickup with a tank on back, followed the driver to his camp 15 miles into the mountains, and sold him the load of fuel. Often, under a contract he had with a Chicago company, he would get up after midnight to fuel trucks passing through to the West Coast.

Osborne visualized every person he came in contact with as a potential customer. His outgoing, congenial personality made him a super salesman. When he drove to Salt Lake City for business or pleasure, he often sold a car to those in his company before they arrived back home. More than once when his wife was shopping, she returned to the parking lot to find to her dismay that Osborne had sold her car. He generally arrived at his dealership at 6:00 a.m. looking to catch farmers before they went to their fields. His associates claimed he sold most of his cars before breakfast.

Ever vigilant, he left few stones unturned to expand his rapidly growing holdings. As he told one of his mechanics, "The dif-

ference between you and me is that if you were running this and
sold a new car and made a good profit, you'd think of taking the
day off. Not me. If there's a guy down the road a few miles with
a flat tire, I'd go down and fix it, even if it only involved making
50 cents."

Osborne's capacity to find new opportunities resembled a
hunting dog's ability to smell a pheasant. How Osborne ultimately became an automobile dealer involved several such inci-
dents. After opening the service station in Soda Springs in the fall
of 1940, he set up a bulk plant and arranged through Reuel to
become a Sinclair jobber. He then started delivering fuel and heating oil to farmers in a radius of 40 miles. When gasoline was
rationed during World War II, Osborne demonstrated Call ingenuity by finding other ways to supplement his income. He first
arranged to deliver freight for the local railroad company and then
a major truck line. As automobile production dropped off and
used cars became scarce, he would buy old taxicabs and other
well-worn vehicles in Salt Lake City, take them to his shop in Soda
Springs for refurbishing, and sell them on his lot next door.

His success with used cars provided the incentive at the war's
end to acquire a Plymouth-DeSoto dealership and one for
Diamond T trucks. To house his new business, Osborne expanded
his service station, erected a small repair shop, and added a showroom for two cars. He kept the Plymouth agency until 1954 when
the Chevrolet-Oldsmobile dealership (including Chevrolet trucks)
became available, a coveted prize since Chevrolet was the highest-
selling car in America. Recognizing its value, Osborne expanded
and updated his showroom and service area, making it the talk of
the small community. With 20 or more vehicles for sale on his out-
door lot and four in the showroom, he was the area's largest
dealer. With Osborne's charm and willingness to bargain, he sold
an astonishing number of cars for a small, rural agency. Being
highly profitable, the dealership became the money machine that
had the potential to fulfill his goal of being a millionaire.

Osborne enjoyed the status of owning the area's premier auto
agency and liked being a salesman. However, for someone with

Osborne's Soda Springs station and dealership. Second South and Main.
(March 1947)

"building" in his blood, being tied to a showroom every day proved too confining. In addition, being aware of Reuel's success with retail service stations, Osborne wanted more of the action, especially since he now had the capital to be aggressive. When an attractive offer surfaced in 1958, he sold the dealership, but kept the Sinclair station and buildings.

Even when the automobile business demanded most of his time and was extremely profitable, Osborne never divorced himself from being a builder. As soon as World War II ended, he was back constructing service stations. He added to his real estate holdings in 1947 by joining with another partner and purchasing a hotel Reuel had built in Afton. The next year Osborne sparkled up the local community by constructing near the station a 32-unit lodge, the area's largest. The motel was an instant success, although within several years others opened and the market became more competitive. In 1960, he constructed a local bowling alley followed by the purchase of a major office building in Kemmerer, Wyoming. Next, he obtained contracts with the federal government to build and lease back post offices in Soda Springs, Ashton, and Gooding, Idaho. Meanwhile, he was increasing rather than diminishing his retail gas business, operating a chain of more than a dozen stations.

This flurry of activity would not be unusual for someone who was sitting back and contracting these projects, but like Thomas (his father), Osborne—driven by frugality—was reluctant to pay others if he could do it himself. He would occasionally get someone to help handle the heavy work on a new station, but he generally ended up doing the painting and finishing touches. His children remember how he dug the sewer and water lines for their new house with a shovel rather than rent a backhoe, and became exhausted from feverishly finishing the cement in the rain.

His strong work ethic might have partially been his downfall. Osborne had a heart attack in 1961 and was told to cut back on physical activity by resting three or four hours a day. He was just 44 at the time and well on his way to being a millionaire by age 50. After his heart attack, he purchased a Rambler dealership intending to work less and spend more time at home. However, he could never resist the urge to pour cement or pick up a paintbrush if he had a project underway. With his go-go mentality, idleness was a sign of waste. Once when fishing on a local reservoir, he grew so anxious from just watching his fishing pole that he insisted on rowing the boat around in circles. A doctor's order to rest each day was like asking someone with a passion for food to stop eating.

In the spring of 1964, Osborne had 19 stations located primarily in Idaho towns such as Pocatello, Burley, Twin Falls, and Boise. Each facility carried the recently established Maverik name with Reuel as the primary wholesale fuel provider. These small cut-rate stations required a typical investment of $7,000 to $12,000, although one reached the hefty sum of $15,000. Osborne's success in retailing gasoline mimicked Reuel's strategy: find locations on major highway arteries on the outskirts of a community where property is less expensive, make the investment as small as possible, operate as an independent to keep fuel costs down, and always price competitively. The key was keeping required capital to a minimum and finding a location where the only competitors were major brand dealers.

In April 1964, Reuel and Osborne decided to combine their petroleum interests and form a corporation—Caribou Four

Corners. The name was selected representing Osborne's Caribou ties (his motel, called the Caribou Lodge, and Caribou County where Soda Springs is located) and Reuel's long-standing holdings in the Four Corners area of Utah, Colorado, New Mexico, and Arizona. Osborne contributed his 19 stations, several trucks, and some land. Reuel put in approximately 50 stations and 20 trucks, a refinery under construction at Farmington, New Mexico, and a building complex including a two-year-old refinery just north of Salt Lake City. Osborne received 19 percent of the stock, Reuel 71 percent, and John Wallace, a Soda Springs banker, 10 percent for $300,000 he invested in the company.

The merger took place for a variety of reasons. Obviously, with their combined holdings, the company became a significant player in the retail petroleum industry in at least a four-state area. By spreading out geographically, they were less vulnerable to a business turndown in one area or the rise of a strong local competitor. Each needed backup since Osborne was in bad health, and Reuel was looking for some assistance in running his operations. And, probably most importantly, they had tremendous respect for each other and assumed that synergy would be created through combining.

The questionable feature of the merger was how they would work together since each was highly independent and liked to run his own show. Unfortunately, Reuel and Osborne were too similar to benefit from possible complementary skills. Both liked to make deals, eagerly searched for new opportunities, resisted constraints imposed by others, and kept their lives private. Neither was content nor necessarily skillful in sitting back and managing once a new business or group of businesses was established.

In September 1964, six months after the merger, Osborne and Jay were planning to meet in Montana to build a station for Caribou Four Corners. Osborne stopped in Pocatello to do some cement work with one of Caribou's employees. Later the two were in Idaho Falls having supper at a restaurant when Osborne walked to the cash register and keeled over from a fatal heart attack. In some respects, those close to him considered it a fulfill-

ment of his wishes since he had once told Jay's wife that he would
rather be dead than unable to work.

Life in a Rural Community

As a result of Osborne's business success, Jay experienced growing
up in a small rural community where initially the family was
impoverished and lived in modest housing, but by the time he was
in high school, they were affluent and resided in fashionable
accommodations. He was raised in a home where his father always
put business first, worked long hours, and left most of the task of
raising the five children to his wife.

Jay's early years were typical of youngsters growing up in a
rural western community. He fished in local streams, built rafts to
float on a nearby reservoir, took long rides on a bicycle, and had a
small dog that was his constant companion. The family had few
vacations, typically involving two days or less. However, when Jay
became of age, he (thanks to his father's dealership) always had
access to a car or motorcycle and ample fuel, and was most content when off on his own. As a youngster, he learned the value of
work. At 14, he qualified for an Idaho driver's license. Soon after,
during summers and on weekends, he delivered fuel to farmers,
some of whom were aghast because the driver of the truck was so
small he had difficulty peering over the steering wheel. In his
teens, Jay pumped gas, helped detail cars, ran errands to get tires
from a supplier, helped sell and install the tires, and even got
involved in talking with customers who were interested in purchasing automobiles.

Jay's outlook changed considerably during his school years.
While attending the local four-year high school, he initially gave
little hint of being a nonconformist. He was elected to several
positions as a class officer, did well in school, and maintained
favorable relations with his teachers, family, and other adult over-
seers. Many were aware that he scored high on intelligence tests,
and he was recognized as a young person with significant potential. Contrary to their expectations, though, the further he moved

Osborne Call family at the time of Jay's high school graduation. Left to right: Craig, Sharon, Janice, Jay, Lance, Osborne, and Candace.

along the more bored he became with school. For someone with a short attention span who was constantly seeking new, exciting experiences, Jay was easily put off by repetition or activity that seemed irrelevant. Listening to lectures and spending time memorizing data to parrot back on examinations did not fit his temperament. He was elected as a junior class officer and maintained average grades, but his primary interests were social—his friends and after-school activities.

Jay was always outgoing, inquisitive, and full of fun. With his charisma, rare gift for conversation, and ready access to an automobile, he had an abundance of boy and girl friends. Early he displayed his fetish for cleanliness. He would polish his cars until they sparkled. If friends were to ride with him, they knew better than to wear dirty clothes or have muddy feet. To escape the dust and the possibility of chipping the paint or windshield, he avoided driving on gravel roads. He would customize his cars and sometimes test their limits in a drag race or by going up the hairpin curves on Teton Pass leading to Jackson Hole, Wyoming. With his constant yearning for adventure, when not required to work Jay liked to get out of town on weekends and evenings and, as one friend put it, "see what was across the next hill." Handsome, exciting to be around, and always on the go, some friends referred to him as the "James Dean of Soda Springs."

Jay's behavior was typical of many teenagers with unusual ability, potential, and zeal. They want to get on with their lives and feel they can be achievers through some route other than schooling. Like Reuel and his father, Jay yearned for independence and control of his destiny. Likewise, he resembled his uncle and father by being strong-willed and determined, and he worked hard at activities he enjoyed. Jay, more so than his father, was often inclined to challenge conventional social and religious norms. Given these similarities, it is not surprising that strained relationships developed between the father and son.

This clash of these two personalities strongly influenced the direction Jay's life would take. As a youth, Jay had his mother as an advocate, although she acknowledges that, like many teenagers, when 16 and 17 he was "difficult for me to handle." Osborne was a stern disciplinarian, much like Thomas, his father. Osborne closely monitored Jay's activities and would rarely back off without some caustic remark. With both father and son being headstrong and proud, neither wanted to give in. Osborne's ironhanded approach caused sparks to fly, often over situations that were of minor consequence. In later years, each realized that their shared traits were the likely cause of their discord.

Although both parties were at fault, Osborne made the tragic mistake that fathers often do with eldest sons. Holding high expectations, the father wants the son to either follow in his path or achieve goals he has failed to attain. When the son shows signs of not moving in the desired direction, the father tends to tighten the reins and to find fault more quickly. To gain the upper hand, the father unknowingly ups his demands and looks for his child to buckle under. In this psychological tug of war, the son is reluctant to lose face. He wants to show he can do it on his own, avoids contact with his antagonist, and finds comfort in his friends. In the father's determination to "teach his son a lesson," confrontation accelerates, animosity grows, and both parties let emotion rather than reason dictate their relationship.

The chasm between Jay and his father did not develop from lack of mutual respect. Osborne was quick to criticize because he thought his talented son was not living up to his expectations. On the other hand, as Jay states, "I had a bitterness toward my dad and yet I had a tremendous respect for him." As a son, he was proud that community business leaders looked up to Oz as one who could "turn a dime into a dollar" and one who demonstrated unusual business acumen and versatility by being successful in gasoline retailing, car dealerships, construction, building and operating hotels and motels, and handling money. Jay often quoted his father when he was out with friends, and he admired his dad's extraordinary drive and way with people, even though their home life was not always pleasant.

Jay's dissatisfaction reached the boiling point from several experiences: In Osborne's efforts to please everyone, he tended to adopt different standards based on who he was associating with at the time. In addition, he was inclined to impose higher work and ethical standards on Jay than he did on himself. It also irritated Jay that Osborne paid him less than other employees doing the same work. Even when he performed well, his father would rarely give him praise. Osborne constantly let others know of his annoyance with youngsters who engaged in such frivolous activities as "joy riding" or staying out late every evening with friends. Later, despite the fact that Jay operated his father's most profitable station, Osborne referred to his son as "one of his worst operators." After Jay acquired a small Tripacer airplane and was landing at the airstrip east of Soda Springs, Osborne commented to one of his neighbors—half in disgust and half in pride—"What am I going to do with that boy?"

While their relationship was often turbulent, one should not read more into this period of their lives than is justified. Few parents of large families escape having one or more of their teenage children suddenly display adverse behavior as they attempt to loosen the reins of parental authority. Theirs was a classic confrontation. Parents want children to carry on family traditions, accept particular values, and become members in good standing of their religious faith. Children are often determined to show they can think and act on their own, enjoy testing social and religious norms, and find reasons for rejecting parental direction. The net result is the same. The combination of a father holding unrealistic expectations for his eldest son, and a son who is confident, capable, and headstrong creates a rift that can take years to overcome.

Jay avoided confrontation with his father by going to the Jackson Hole area of Wyoming during the summer between his junior and senior high school years, where he found employment at Jackson Lake Lodge as a dishwasher. The next summer, after graduating from high school, he worked in his father's businesses. In the fall he enrolled at Brigham Young University and essentially wasted nine months since his motive was not to learn but to get out of Soda Springs, and besides, he had few other viable opportunities at the time.

The following summer, Jay returned to the Jackson area, this time to manage a station for his father. This was Osborne's way of keeping in close contact while simultaneously offering Jay a maturing opportunity. Osborne knew he was placing substantial responsibility on someone not out of his teens and showing little sign of settling down.

Osborne named the Jackson station "Jays CutRate." It consisted of a 15-by-25-foot frame structure with four pumps in front. When having breakfast at a bakery close by, Jay met a young waitress, Teddy Lou Brown, who changed his plans for the coming year. In the fall, she planned to return to her home ten miles west of Rexburg, Idaho, and complete her senior year at the local high school. After frequent dates in the summer, Jay decided to enroll at Ricks College, also in Rexburg, rather than go back to BYU. He lacked the option of continuing to work at the station since Osborne closed it during the winter. At the time, skiing had not yet gained sufficient popularity to make Jackson a year-round resort.

The activity in Rexburg that dominated Jay's time and proved crucial to his future was his evolving relationship with Teddy. They

Jackson Hole, Wyoming, station that Jay managed for his father in the summer of 1959.

were engaged at Christmas and married June 1, 1960, after her graduation. Both sets of parents thought Jay and Teddy were too young for marriage. However, the Calls assumed marriage would settle Jay down, and they knew he lacked interest in continuing his education. The young couple had little time for a honeymoon since Jay had agreed to return to operate the Jackson station. As a wedding present, Jay's parents gave them a used 23-foot house trailer that served as their living quarters next to the station. Jay spent many hours sanding and painting his new home. Their finances were such that Teddy took the remains of the $50 her parents had given them for their honeymoon, obtained small change at the bakery, and crossed her fingers hoping that their first customers would not come in with large bills.

Jay and Osborne's agreement was to split the station's net profits. Given such an incentive, Jay quickly responded to the challenge. Both he and Teddy worked long hours during a record tourist season, and when it came time to close in the fall their share was $2,000 (equivalent to more than $16,000 today). After noting Jay's success in Jackson, Osborne urged his son to operate his new but failing outlet south of Willard, Utah, on Highway 89 between Ogden and Brigham City. (Interstate Highway 15 later bypassed this section of the highway.) The manager who operated the station for the first six months achieved only modest sales, and Osborne let him go, hoping that Jay could move in and turn the business around.

Thus, in the fall of 1960, Jay took over the small, modern-appearing Maverik station with four pumps in front and began searching for ways to expand sales. His first action was to improve customer service. In this pre–self service era, Jay would fill the customer's tank, check the oil, and wash the windows, all on the run. He was fast, businesslike, and rarely had time to chat or be friendly. His hours were 7:00 a.m. to 10:00 p.m., and for the first nine months, he never had a full-time employee. As he states, "For the first 30 days I never left the lot." Jay had the well-known retail "B shift": Be there in the morning, be there all day, and be there at night.

One of the few local businesses was a trucking company. Jay offered the owner attractive rates if he would buy his fuel from Maverik. The person signed on but only if he could gas up at night since his current supplier north of Willard stayed open 24 hours. Jay agreed although he and his family never became accustomed to the blast of the air horn at 2:00 a.m. as they slept in the trailer south of the station.

Even with business increasing, sales during the first winter never exceeded a daily rate of 500 gallons. Thus, Jay and his family lived on a meager income since Osborne (in typical Call fashion) did not provide Jay a working capital advance or other financial support. Jay's family was always glad to see Teddy's parents arrive from their Idaho farm because they brought beef, potatoes, and other eatables.

Eventually the young couple's work paid off. By midsummer, sales had more than doubled. Jay still recalls the thrill of his first 1,000-gallon day. When the station's daily average reached 1,500 gallons, he hired additional help, but this did little to prevent him from being constantly on the run. They lived by the station until Teddy was pregnant with their second child, after which they moved into a larger mobile home.

When sales eventually tripled, Jay kept raising the bar of what he wanted to attain. Fortunately, Osborne had taught him that money could be made in many ways if one can identify the opportunities. Wanting to take advantage of his selling experience, Jay went searching for financing to open a used-car lot next to his station. After his current Brigham City banker refused to lend him the necessary $3,000, Jay went to the other local bank hoping that his enthusiasm would overcome his youth and lack of collateral. Being somewhat small in stature and appearing even younger than his age, Jay boldly asked to talk to the president. (Jay was always clean and neat at the time but not well dressed. He could not even claim ownership to a suit!) The president said that Jay "looked like a 14-year-old kid," but "he had a determined look on his face so I invited him into my office." After listening to his story, the banker consented, stating, "I think we can figure out how we can get you a couple of used cars down there." Bursting with pride, Jay was starting to feel that he could cut the strings from his father.

Jay traveled as far as Denver to get the type of vehicle he wanted. His plan was to appeal to young buyers by obtaining four-speed V-8 sports coupes. Based on prior experience, he detailed and cleaned the cars until they sparkled. He kept three or four in stock and had constant turnover. Like his father, Jay is a super salesman, and at the time he had special appeal because he was the same age as his typical customer.

Soon a related opportunity surfaced that numerous others had rejected. A major car accessories supplier was dropping out of the tire business and wanted to dispose of three to four hundred recap snow tires all in one size, 7.50 by 14. The company representative insisted on selling the entire lot at the bargain-basement price of $4.50 each, but the offer was still unattractive, especially to a small station owner who sold at most a dozen or so tires a month. Jay had everything going against him: the 7.50 by 14 tire did not originate until 1957, and new car owners rarely bought recaps; any experienced dealer would consider it foolhardy to open a snow-tire business with only one size; his station lacked an

Original Maverik station Jay managed and later owned in Willard, Utah.
Jay and his family lived in the trailer to the right.

indoor or covered bay; he did not even own a wheel balancer; and his only jack was a small one designed to fit under a bumper.

Much like when his Uncle Reuel built a high-altitude air-plane, Jay turned what others thought was unthinkable into a business opportunity. First, he got the sellers to accept 90-day terms. Then he placed a large row of tires in clear view in front of the station with a chain wrapped through for security. He adver-tised the tires at the extremely low price of $20 a pair installed. Probably the most ingenious marketing aspect of his offer was an unconditional guarantee if the buyer purchased tubes with the tires. Selling tubes gave him an additional $2 to $3 profit per tire. Of more importance, his dealership quickly gained credence with the local populace when he replaced returned tires on a "no ques-tions asked" basis.

The final ingredient making the venture a success was one every entrepreneur invariably relies on: luck. Winter came early and it came hard. After a six-to-eight-inch snowfall in September, most dealers were out of snow tires. Dealers as well as dozens of retail customers clamored to buy them from Jay. He paid back the $1,800 to his supplier before the 90-day deadline and had considerable cash in reserve plus the remaining inventory. He used the cash to broaden his tire line by offering snow tires in different sizes. Ultimately, tires became a major source of his income. Just prior to his father's death, Jay, ever on the lookout for new opportunities, was considering putting in a large recap plant near the station.

After expanding into cars and tires, Jay even became involved in a partnership to mix and sell fertilizer, the only one of his early ventures that failed. However, as happens with most entrepreneurs, it taught him a valuable lesson: to be cautious in moving out of his field of expertise.

Within three years, Jay had gained his father's respect by making the Willard station Osborne's largest volume outlet. Like many of his relatives, this success only whetted Jay's appetite to be on his own. When he began looking for property, Osborne, reluctant to see him leave, offered Jay the Willard station at fair market value under the stipulation that he would pay it off within several years. After Jay agreed, they jointly purchased a tanker truck, and the two previous antagonists were now considering building several stations together.

At this time, the Maverik stations owned by Osborne and Reuel were managed under a system developed by Reuel wherein the operator was in effect a consignee. The landlord owned the fuel inventory as well as the station, and the operator was charged a set amount per gallon. Any revenue above that could be retained with the understanding that the operator paid all expenses other than fuel. In addition, the operator kept all income from nonfuel merchandise, which prior to the rise of convenience stores was rather meager. This system greatly simplified the financial relationship between the owner and dealer. The only accounting necessary was to keep track of the gallons sold and maintain a record of the daily bank deposits the dealer made in the owner's name. This arrangement proved extremely effective, one Jay later adopted with his station managers.

Working under this arrangement proved to be contentious to both Jay and his father. Under the agreement, with the operator's

Osborne and Jay by the fuel truck they jointly owned.

primary concern being the margin between the amount per gallon
charged by the owner and the retail price, Jay became incensed
when Osborne initially gave him four cents compared to five cents
for other operators. Later Jay would bristle when his father came
around if he assumed his purpose was to give advice. Once when
Osborne brought a friend to demonstrate how a $50 punch card
could be used to attract customers, Jay quickly let both his father
and his companion know that he did not need their assistance. In
the youthful urge to demonstrate autonomy, Jay was occasionally
brash and impetuous. However, this same determination and self-
reliance eventually pushed him into the forefront of the oil and
hospitality industry.

After two years of managing his own station, Jay now had that
special feel of success that spurs a person on. He was earning
$20,000 a year, significant income then for a person without a col-
lege degree, and he was starting to build equity in his business.
However, the family's living conditions were far from opulent.
They resided in the mobile home for four years before purchasing

a house in Brigham City (an obvious carryover of the Thomas Call "frugality" tradition). Like Reuel, Jay's only luxury was an airplane, the Tripacer he acquired in 1962. Incidentally, soon after obtaining his license, one of his initial solo flights almost resulted in his demise. When flying over a mountain valley 30 miles northeast of Brigham City, the engine blew a piston. Fortunately, he was close by a local airport, providing a haven for him to glide in safely.

Insights into Business and Entrepreneurship

Jay's early business endeavors provide useful insights into entrepreneurship. At an early age, he displayed the same determination, action orientation, and strong work ethic that characterize him today. Most of his high school peers assumed he would be an adroit businessman like his father. A few viewed him as being on the fast track to being a millionaire at an early age. One perceptive classmate wrote in his high school yearbook, "Jay, you have a personality that will really take you a long way in this world and win you many friends. You have a way of doing things that no on else could think of." This comment identifies a characteristic that many of Jay's employees and associates consider his key attribute. As one business expert observed, outstanding entrepreneurs are able to "see things that escape others." Jay has this knack of being able to identify business opportunities, evaluate risks, and break complex issues down into a few fundamentals for easy analysis. Most of his associates are amazed at how quickly he gets to the heart of complex issues, and how readily he picks opportunities out of situations that appear ordinary to everyone else.

Another critical skill evident in Jay's early development was his unusual capacity to be open, learn from experience, and change his behavior accordingly. As we shall see, the occasional questionable conduct that marked him as a youth soon disappeared as he matured and accepted other values. His daring, "push to the edge" behavior when behind the wheel of a car was replaced within a few years by a pilot who is extremely cautious when at the controls of an airplane. He became so well trained and orderly that control tower operators have complimented him for his superb approaches and flying ability. His desire to control and occasionally intimidate in time became replaced by the exact opposites—expecting others (both his children and employees) to take initiative, make decisions on their own, and be responsible for the results. His occasionally self-serving behavior was altered by what became a favorite motto: "Always leave others better off from having associated with you." His early reluctance to take advice from adults was soon supplanted by a willingness to listen and to actively seek the counsel of others. Rather than being self-centered, he gains satisfaction out of seeing subordinates grow, and he goes out of his way to shun the spotlight. Many of the above are simply signs of a teenager maturing, but his capacity to adapt quickly to new conditions is an attribute that few people display and has served him well as he broadened his business horizons. (Osborne also learned from the conflict with his son. He was not nearly as demanding or critical with Jay's two younger brothers.)

Jay's relationship with his father supports many of the conclusions reached by experts on entrepreneurship. Warren Bennis, an acknowledged authority on leadership, found in his studies that leaders often "have a strong, determined set of parents" and someone in the family who gave them insights on how to proceed in their careers. In another study by two Syracuse professors, it was found that parental example is more important than parental wealth in determining whether offspring become entrepreneurs. While Osborne and Jay were often at odds, Osborne was an important mentor. He implanted in his son an entrepreneurial mind-set, and he taught him the essence of deal making. Equally important, at an early age Osborne set the stage for Jay to gain confidence by being involved in significant business transactions. Jay still remembers when, at age 11, he sold a customer tires and, at 14, he took the lead in selling another person a car. When 18, making $2,000 during the summer at Jackson went a long way toward boosting his aspirations. With Osborne and Reuel as role models and with the early opportunity for hands-on experience, he had a jump start to fulfilling his entrepreneurial ambitions.

Many conclude that Jay's drive comes from his determination to surpass what both Reuel and his dad had accomplished. Obviously, he did not want to fail in comparison, but he is not one to keep score in such a rivalry. Others suggest that Jay's, Reuel's, and to some degree Osborne's willingness to break from their religion and invite the scorn of friends and relatives are related to their same willingness to take business risks and not be guided by mass mentality, both key essentials in entrepreneurship. As one member of the Call family jokingly stated, "The Calls—some of us have faith and others have money."

Going back to Jay's youth, he always wanted to make his mark in business. However, he did not necessarily assume this would occur through fuel retailing or other enterprises his father had been engaged in. Difficulties with his dad partially soured him on such prospects. Like most entrepreneurs, he was simply looking for opportunities—opportunities that eventually evolved through his knowledge of the retail petroleum business. These experiences started Jay in the direction that would ultimately lead to his incredible rise as one of the most successful current-day entrepreneurs in an industry considered mature, lacking in innovation, and devoid of growth prospects.

STRIKING OUT
ON HIS OWN

"Fortune is not on the side of the faint-hearted."
—Sophocles

With Osborne's death, the fledgling Caribou Four Corners corporation faced a management crisis. The combined company had been operating for just six months, and the retail end of the business was still being consolidated. With two refineries coming on stream, an expanding trucking and wholesaling operation, and more than 60 retail outlets, the firm was rapidly becoming a fully integrated petroleum enterprise and needed more management talent. Reuel's three sons were attending college or teaching school, and Jay, just 24, had limited experience. Even more awkward, the employees providing key administrative support were adequate for a small business but lacked the background and depth to accommodate Caribou's growing, diverse operations.

Osborne's death not only took an emotional toll on his family, but his widow now had to assume the responsibility for looking after family interests in Caribou Four Corners and managing Osborne's other business holdings (the motel, rental properties,

Soda Springs station, etc.). Of Janice's five children, only Jay and his older sister were married. Jay's younger sister was 18 and a recent high school graduate, and the two younger boys were 16 and 11. With three children at home and a variety of business interests to look after, Janice's problems were considerably more than coping with the emotional loss of a spouse.

When Osborne died, Reuel's goal was to keep the company intact. Its sheer size was impressive for a local private corporation and the merger offered numerous economies of scale. For the corporation to pay off Osborne's family was out of the question since Reuel continually pressed the limits of his financial credit, and the company was woefully undercapitalized. Thus, the new corporation could not buy back even a small portion of the shares held by Janice and her family.

Reuel, wanting to keep the company together, hoped he could convince Jay to come aboard as a vice president. At this point, Reuel knew Jay primarily through his dealings with Osborne and was aware of Jay's success at the Willard station. As Reuel became more familiar with his nephew, he considered him to be confident, aggressive, inquisitive, and a quick learner. Of more importance, he was a self-starter and willing to make the personal sacrifices necessary to ramrod new projects into being. In addition, their shared love of flying and similar philosophies of work and life led to a growing bonding.

When Jay received the offer to join Caribou, he was not sure how to respond. At the time Caribou Four Corners was originally formed, he had turned down an invitation to put his Willard station into the company, and he had plans underway to add more stations on his own. However, he welcomed the prospect of further association with his favorite uncle, and he knew he would benefit from experience in a larger company. Urged on by his mother, he accepted the offer in November 1964.

Consistent with his management style, Reuel brought Jay aboard with no specific assignment. His initial responsibility was to finish several projects Osborne had initiated. Later, he spent most of his time in building and overseeing stations. In addition,

he worked on collecting major overdue accounts, helped fabri-
cate highway signs to advertise locations, and actively marketed
Caribou fuel to new wholesale customers. On occasion, he
served as a troubleshooter and point man on more difficult problems.

Reuel, eager to keep Jay on board, knew which incentives to
dangle in front of his nephew. Reuel offered to provide him a
Mooney airplane (a popular single-engine, four-passenger aircraft
that cruises at 180 MPH) if he collected $30,000 from a delin-
quent account in Idaho. Reuel's offer was also made out of concern for Jay's safety; Reuel did not like Jay flying the single-engine
Tripacer, a plane with a high fabric-covered wing that he had salvaged out of a farmer's field in Colorado. Driven by an incentive
similar to letting hunting dogs loose after a cougar, Jay was soon
flying the Mooney in what he described as "a red-letter day for
me."

Jay stayed with Caribou Four Corners for more than three
years (until February 1968). The primary motive for his eventual
resignation was the familiar Call incentive—he wanted to run his
own show. As much as he admired Reuel, Jay would not be content until he was his own boss. Osborne would have admired his
son's decision. Salary was not a factor. Jay's pay was adequate
though by no means excessive. Individuals associated with
Caribou at the time, such as Mel Baird, were not surprised at his
decision. Mel knew that "right from the start Jay was going to do
it." On one occasion, Jay had revealed his "all or nothing" attitude by telling Reuel, "I am either going to create a big business
or be a clerk in a Holiday Inn." Reuel saw so much of himself in
Jay that he made little attempt to deter him, although prior to
leaving, Jay had to apply pressure to gain approval to establish his
own stations in Ontario, Oregon (just across the Idaho border);
Lewiston, Idaho; and Richland, Washington.

Jay did have some concerns about Caribou Four Corners. On
occasion, he could not understand why Reuel was not more
aggressive, especially in using long-term debt to finance expansion. In addition, one of Reuel's sons was scheduled to return to

Jay standing by his Tripacer airplane, 1962.

Reuel and Jay in front of a Caribou Four Corners's airplane (1966).

Caribou as a vice president in late 1966, and the other two sons gave indications of likely joining, which they did two years later. Management would soon be crowded with Reuel's offspring, and Janice and her family had limited power as minority stockholders. In addition, some contention was developing between Osborne's heirs and the Caribou Four Corners board of directors, and Jay wanted to avoid being caught in the middle.

As time went on, Jay's mother became more dissatisfied with being unable to get a cash return on the Caribou stock owned by her family. She still had three children at home, and most of Osborne's assets were in real estate, not cash. In Osborne's brief legal will, he followed the typical practice of leaving half his estate to his wife and the balance split among the children. At the time of his death, only a small distribution was made to the siblings. Approximately half of Osborne's assets were tied up in Caribou Four Corners and the other half in the Caribou Lodge, post offices, the building in Kemmerer, and other property. The bulk of the non–Caribou Four Corners assets was placed in a family partnership that is still in existence based on the guidelines contained in the will.

Thus, Janice began pushing for greater income from Caribou's board. The board continually rejected proposals to buy back stock or pay dividends since the company was asset rich and cash poor. In addition, because it was a private not a public corporation, the stock could not easily be sold externally. It was more than a dozen years before Osborne's family members received significant income from their Caribou shares. Then the $1.5 million received was staggered over several years. Since this did not occur until the late 1970s, Jay's rapidly growing business empire was not funded by his family's wealth. By 1977, he had 50 stations and already exceeded his father's fortune. The common misconception that Jay was bankrolled by family money is nullified by Fred Baugh, the Brigham City CPA who put together Flying J's annual financial statements. Baugh states, "I never did see any influx of capital from his dad or any other family source."

Early Additions to Jay's Service Station Holdings

In the spring of 1968, fortified by his experiences in working for Caribou Four Corners and operating several stations of his own, Jay was poised to pursue his goal of being a highly successful businessman by the age of 30. Jay learned important lessons through establishing his initial stations. To finance the Ontario station built in 1965, he obtained a $10,000 unsecured loan from the Box Elder County Bank. By making a small down payment, he acquired the land for $10,000, after which he added an attractive, spacious building for another $17,000. Besides fuel, he sold groceries (becoming a forerunner of the C-store) primarily due to an Oregon law that restricted gasoline stations from being self-service. This requirement forced most stations to have a minimum of two attendants, one of whom handled the cash register with spare time to ring up groceries.

The Lewiston station, built in 1967, had four pumps with a trailer house as a pay booth. Being self-service, this kept Jay's investment and operating expenses lower. The innovation he introduced at this location was to use the trailer house as a live-in facility for the operators, a concept adopted by several competitors and one that Jay used to advantage when he initially lived next to the Willard station. Using Reuel's consignment scheme, Jay did not hire the station operators as employees. Jay owned the facility and fuel inventory, and the operators received free rent and a set amount on each gallon sold. The primary advantages of this arrangement were greater security from having someone on-site day and night, and the operators had more incentive to treat customers right and to push sales.

The year the current Flying J corporation came into being was 1968 when, soon after leaving Caribou, Jay dropped the Maverik title, incorporated as Flying J Inc., and renamed his stations Fastway. With the Willard outlet, Jay had four stations at the time. Most histories date the company from 1968, although Jay had owned stations seven years earlier. People familiar with Jay generally assume the corporate designation came from his love of

Lewiston, Idaho, station (1968)

Barstow, California, station (1968)

flying. However, he viewed the title more as a cattle brand or a name someone might give to a ranch. It was a decade later before he switched his small gas facilities to the Flying J flag.

The Fastway outlet in Richland, Washington, (finished in 1968) was the same design as the one in Ontario except the grocery space was converted to a live-in facility for the operators. The initial results at Richland even astounded Jay. Projected sales of 40,000 to 50,000 gallons per month (twice that of an average brand station) were eclipsed within the first 30 days. Sales skyrocketed to more than 180,000 gallons monthly, resulting in his first retail gusher since Willard. The undersized facility was often clogged with long lines of cars waiting at the pumps. Such a spectacle fueled Jay's entrepreneurial ambitions, later acknowledging that it "really got me going."

Budding entrepreneurs discover that the first ventures are generally the most difficult to finance and make profitable. Likewise, they typically provide the best learning experiences. Although the Ontario site was "not the best" by Jay's own admission, he found that with only a $10,000 loan and a small amount of personal capital, he could build a station and pay for it from cash flow. In addition, he discovered that he could turn his fuel suppliers into cash cows. At the time, a dealer received a 1 percent discount if the bill was paid within ten days. Fortunately, Jay's fuel suppliers (primarily Maverik) were slow in billing. Reuel's loose administration was now benefiting Jay. The billing lag time often exceeded two or three weeks, giving Jay up to 30 days after receiving the fuel to make payment and still receive the 1 percent reduction. Flying J's low fuel prices generated huge volumes that resulted in quick inventory turnover. Thus, Jay would sell one or two loads before paying the supplier. In the example of Richland, with monthly sales of 200,000 gallons priced at 30 cents a gallon, he received $60,000 in fuel receipts before he had to pay for fuel already sold. The $60,000 was more than the total cost of the station, so Jay was expanding on his suppliers' money.

With this excessive cash flow (even after making fuel and major mortgage payments), Jay had enough left to start another

Fresno, California, station (1970)

station. Accountants advised him of the extreme danger in funding capital facilities from cash flow, but early on Jay displayed his almost complete disregard for risk when he had confidence in his decisions. As time went on, he was careful to pay off several stations, giving him the collateral to back a loan if necessary.

With his financing in place and the advantage gained through live-in operators, Jay was positioned to press forward with his expansionist plans. Between 1968 and 1973, he built 23 stations, 9 in California, 8 in Washington, 5 in Oregon, and 1 in Nevada. Except for Portland where he had three stations, most were in what then were medium-sized communities, such as Fresno and Bakersfield, or in smaller towns like Petaluma, Eureka, Walla Walla, and Klamath Falls. The common feature that drew him to these communities was that they all lacked cut-rate stations. His initial marketing strategy mimicked his fathers and Reuel's: minimize the investment, maintain lower prices, undercut the majors, and keep to rural locations.

Jay's decision to concentrate on the Northwest and California was prompted by several other factors. Initially he deemed it improper to compete with Reuel, who had Maverik stores in Idaho, Utah, and Wyoming. In fact, Jay did not build a station in these three states for six years (the first in 1973, when he constructed a C-store in Brigham City on the Main Street edge of the site he had selected for Flying J corporate headquarters). Jay also chose to concentrate on the states west of Utah that were slow to adopt cut-rate stations. The three-state area Maverik operated in was the hotbed in the West for this type of facility. Upwards of 40 percent of Utah stations were self-service, whereas Washington and California had less than 10 percent. The West Coast was dominated by major-brand dealers who, at the time, charged 37 cents a gallon for regular gasoline. (The wholesale cost of fuel plus federal and state taxes generally accounted for 25 cents with a markup of 12 cents). Jay marked up fuel half that of branded dealers, and, as a result, he sold at least three times their average volume, making him more profitable. As Lee Liberatore, one of Flying J's early Washington employees, said, "When competitors found out how much gallonage we pumped, they said we were liars and must be doctoring the figures until they sat across the street and counted the customers."

Jay's decision to spread his operations along the West Coast made sense in two other ways. With limited equity, he knew that by concentrating in one area a local gas war could wipe him out. Second, dispersed operations were less of a hindrance to him. By air, he could reach his West Coast outlets in one-fourth the time of a competitor traveling by car.

As a bonus, Jay could engage in his favorite activity—flying. When he left Caribou in 1968, he purchased the company's Mooney he had been using. Soon after, he replaced it with a Beechcraft Bonanza, a larger, more luxurious, five-place airplane. Within 18 months he acquired an Aerostar—a six-place, twin-engine, turbocharged aircraft that requires the pilot to have multi-engine instrument ratings. The casual observer might conclude that Jay squandered funds to satisfy his flying fancy, but those close to him knew the benefits he derived from flying. As Bob Smith, an Oregon real estate developer and close friend, stated, "Most people do not know how to use an airplane. With most businesses it's a cost. With Jay it's one of his greatest assets."

In time, Jay revised his business strategy by placing even greater emphasis on volume. This required locating in areas with higher traffic flows. Generally, rural sites obtain abnormally high volumes only if they are strategically located on a main traffic artery and have few cut-rate stations as competitors. Property in densely populated communities is more expensive and the facilities more costly, but the potential volume can offset higher capital investments.

The first major test of this revised strategy occurred in Monterey, California, in 1970. In a "go for it all" gamble, Jay obtained property in the city just off the freeway for $100,000 (by far the largest amount he had yet paid for unimproved real estate). After a long struggle to get the appropriate government approvals, he constructed a 12-pump station and live-in facility. In total, the investment represented 90 percent of the company's equity. Fortunately, the design of the plaza, aided by attractive landscaping, gave the station enormous visual appeal and helped turn it into an instant financial bonanza. It was the company's crown jewel for nearly a decade. As Ron Brisendine, Flying J's first manager over retail operations, stated, "At Monterey, Jay was initially way over his head. Fortunately sales started off at more than 200,000 gallons a month, and the station generally led the company in sales until the first truck plaza." No longer wary of expensive locations, sites large enough to accommodate a motel as well as a station were soon added in Reno, Nevada, and other locations.

Building Staff at the Corporate Office

Each week Jay was either in the air or on the road from Tuesday through Friday visiting stations, looking for sites, and engaging in related business. He was in the corporate office only on Mondays and occasionally on weekends. The operators anxiously awaited visits from the enterprising, personable owner although some were taken aback when he arrived in a black leather coat on a motorcycle. Once when making the rounds by motorcycle, he stood on

Fastway station in Monterey, California, 1970.

the perimeter of a Flying J property viewing the facilities. An attendant who had never met him was about to call the police when another employee identified the suspicious man as the owner.

Many wondered how Flying J's business could prosper when the chief executive was rarely there. Rather than being a problem this aloof management approach was part of Jay's strategy. By being away, he forced subordinates, many with limited experience, to take responsibility and make decisions.

Selecting competent subordinates is often the difference between success and failure for a first-time entrepreneur. It was a skill Jay developed over time. Initially he employed many of his Soda Springs High School friends. Some turned out to be extremely valuable, but others proved disappointing and soon left. When recruiting, he obtained references from business acquaintances and others, but being a sound judge of people, he mostly relied on his own assessments. Like his father, he has the skill to see through facades put up by manipulators. One technique he frequently used was to walk candidates to their cars after an interview. He calculated that if a person's car was dirty on the outside and cluttered inside, this is how the individual would maintain a station or office. Early he developed the conviction that the best gauge of a new employee was to give the person a relatively open-ended job description and see how effectively the individual could pick up what needed to be accomplished.

Four employees played primary roles in the company's early development. The first was Marcella Hume, Flying J's initial full-time office help. She was hired in 1969 as bookkeeper and later became the office manager. With Jay gone most of the time, Marcella became the mother hen of the company and assumed a major role in coordinating activities at the corporate office, especially personnel matters. Marcella had a keen awareness of the goings on within the home office and in the field. Each Monday morning Jay would listen to her recount the happenings of the prior week.

Another valuable recruit was Ron Brisendine, a retailing specialist. Ron attended high school in Soda Springs and worked in retail grocery stores before managing Jay's station in Ontario, Oregon. He left Flying J, returned to the grocery business for two years, and was then rehired to operate the new Fastway station in Vancouver, Washington. The company also had plans to build two stations near Portland when the manager over construction became ill. Soon after, when Jay came to visit, Ron volunteered to oversee the construction although he had no experience to justify such an offer. Relying on Ron's innate ability, Jay reached in his airplane, handed him the plans, and said "Good luck." As Ron noted, "Not many people could do that," adding, "One of the key things leading to Jay's success is his ability to let somebody else do stuff without hounding him." Jay placed such confidence in the people he worked with that they were determined not to let him down.

After Ron showed this initiative, Jay made him manager over the Northwest stations and two years later moved him back to Brigham City as director of marketing, a position involving responsibility for all retail operations. Ron was extremely likeable and a key figure in the "work hard, play hard" clan surrounding Jay that devoted almost their entire effort to the company. When Ron left in 1985, retail sales had been growing at more than 25 percent per year (due in no small part to his ability and commitment) and the company now had 55 stations that were the envy of most competitors.

The next employee to fill an important initial role was Richard E. (universally known as "Buzz") Germer. After finishing college in 1972, Buzz joined the company and was soon appointed manager over California retail operations. Buzz and his family became some of Jay's closest friends. Buzz eventually became an expert in refining, supply, and transportation—functions paramount to the rise of Flying J. Currently, he is president of Big West Oil Company (a wholly owned subsidiary of Flying J), a senior vice president, and on Flying J's board of directors.

The way Buzz was hired and trained reflects Jay's unusual management approach. Just prior to Buzz's graduating from Utah State University with a degree in business, Jay arranged through a friend for an interview. Buzz had little interest in working in a service station, but he was enthusiastic about Jay's plans and attracted by his upbeat personality. Jay indicated he needed help in station management and construction, but said little else. After Buzz agreed to come to work, Jay assigned him to manage the nine California stations and oversee construction in that area.

A few days later, Jay and Buzz flew to Los Angeles for what Buzz assumed was an orientation tour. They rented a car and drove to the Fastway station in Bakersfield. Jay told Buzz that he could live in Bakersfield or at any other southern California location. Jay noted that the Bakersfield facility needed a car wash, and made a few other suggestions before they returned to the airport. Buzz kept waiting for Jay to outline the training program he had in mind. At the airport, Buzz was caught short when Jay took out his suitcase and said he was leaving. Buzz, still in disbelief, asked, "What am I supposed to do?" As Jay departed, he smiled, looked him in the eye and said, "Just learn the business. Give me a call if you need me."

For one stressful month, Buzz heard nothing from Jay. Then one day he flew in and took Buzz to lunch. Their conversation was more social than business. All along Jay had assumed that Buzz was competent to handle the job, and if he needed help, he would ask for it. This story brings out the only complaint Flying J employees occasionally express about their boss: he is sometimes too indirect or waits too long to provide guidance or criticism. On the other hand, Jay is convinced that employees will more rapidly develop their skills and attain their full potential if left to operate on their own.

Jay's early personnel selections were not always on target. He had unusual difficulty in selecting and retaining competent controllers. The first controller left under duress within a short period. Then, Paul Brown, a CPA of unusual ability, took over in 1973 when the company had 20 stations. He arranged for the first Flying J audit and inserted controls over the retail outlets that improved accountability and increased profits. Later, he was instrumental in helping obtain the financing to make the acquisitions that multiplied the company's size on each occasion. Paul left Flying J twice. Each time he was hired back. In between, the company's financial management suffered.

Exploring Partnerships

Starting with an airplane purchase in 1962, Jay experimented with partnerships. Later he had partners in other airplanes, motorcycles, a television cable company, gas stations, a truck plaza, apartments, restaurants, the fertilizer business, and Green Thumb, a company that sold plants. His largest single joint venture was with Maynard Victor, an acquaintance from Brigham City. They had two filling stations in California, a station and car wash in Ogden, an airplane, motorcycles, and property including restaurants in Salt Lake City. Some ventures were set up as a way of helping others, such as the Alpine Oil Company with his brother Craig, or they were personal projects he wanted to keep separate from Flying J. Occasionally they were established to avoid overburdening his company's already lopsided balance sheet.

Like many entrepreneurs, Jay found he was more effective operating on his own, and by 1976 he had terminated any remaining joint agreements. His reasoning was that he did not have time to keep on top of each venture or, opposingly, that he was running them with limited support from the other party. Partnerships that prospered provided few problems, but the troubled ones frustrated him because he lacked control. And, as he stated, "When the wolves approach, partners fight," which was never to his liking. However, Jay is not one to be bitter over relationships in a business failure or dispute. Once when he lost $300,000 in a venture where he had little input, he gracefully backed out, telling the partner and especially the partner's father that their friendship was more important than the money he lost. According to Jay's close associates, they have never heard him "bad-mouth" anyone, even those who absconded with his money.

Overcoming the Financing Hurdles

Bankers, financiers, and competitors have been awed by the way Jay obtained the financing to build such a competitive group of stations. Within five years, he had parlayed 4 facilities into 27. At the end of Flying J's 1973 fiscal year, annual corporate sales had increased 13 times (from $736,000 to $10,039,000). Profits were up nine times ($35,000 to $316,000). Flying J owners' equity, pushed by an unheard of average annual return on equity of more than 50 percent, had jumped from $81,000 to $604,000—a gain beyond the dreams of all but the most optimistic investors. Jay's average station was making a monthly profit of nearly $1,000, he was adding five outlets annually, and nothing on the horizon stood in the way of his becoming a millionaire and continuing his heady advance.

The way Jay arranged the company's financing provides insight into his personality. The most underestimated element in the typical entrepreneur's business plan is required capital. Underfinancing is the villain that eventually drags under most start-up ventures. Even close observers of Flying J could not understand where Jay obtained his financing. Magazine writers speculated that he had silent partners—either his mother, Reuel, Maverik, or even the Mormon Church. Yet the only contributor of the group was Maverik due to its slow billing, and this was unintended. As noted, in the first two years Jay required little financial backing other than a variety of small, unsecured loans due to generous cash flow from his stations. When this financing became inadequate, he obtained larger loans through a family friend at the First National Bank of Kemmerer (the Wyoming banker who acted as the white knight to save Reuel on several occasions) and from Brigham City's Box Elder County Bank. For these banks, even a $100,000 loan pressed their limits, although once the Kemmerer bank issued Jay an unsecured loan several times larger. By 1970, the Box Elder County Bank was able to satisfy Jay's growing financial appetite by obtaining larger Small Business Administration–backed loans. These more sizable borrowings were feasible because under the SBA program the bank had to guarantee just 10 percent of the total.

In the first few years—to make the balance sheet look more liquid (short-term assets in relation to short-term liabilities)—Jay, at the end of an accounting period, would occasionally get the Wyoming banker to make him a long-term loan and thus build up the cash portion of Flying J's balance sheet. The loan was normally made without collateral, something that always had bank examiners and the bank board questioning Jay's friendly banker's judgment. Jay would pay the loan off within a few months, thus drawing down his cash and, in effect, making the loan a current rather than long-term liability. It improved the appearance of his balance sheet, but bankers continued to overlook his precarious financial position and issued Flying J credit based mainly on the strength of Jay's character and past performance.

Not being shy about walking a financial tightrope, Jay was constantly in trouble with bankers for the company's deplorable debt-to-equity ratio. Bankers prefer a ratio of no more than one-to-one. In 1971, Flying J's shot up to nearly $4 in long-term debt to $1 in equity. This ratio dropped to two-to-one the next year after profits more than doubled to $316,000 although company long-term debt exceeded $1 million for the first time. To get on safer ground and alleviate the problem of constantly monitoring the company's cash position, the next year Jay was able to obtain

a $300,000 Small Business Administration long-term loan
through the Box Elder County Bank. He hoped, at least for now,
that bankers would stop wincing when they reviewed his financial
statements.

The following year Jay turned to Utah's First Security Bank
for financing when the Box Elder County Bank lacked the
resources to meet his needs. As Bob Heiner, former president of
First Security, explained,

> Jay was not an easy person for a banker to handle
> with all of this enthusiasm and confidence.
> Though Flying J's financial position would not
> justify it, it was difficult to turn Jay down based on
> his record. Every time he made the commitment,
> "If you lend me the bucks I can make it work," he
> came through on his promise.

Through this period, even with Flying J's high debt load and substantial interest payments, the company still earned an impressive
10 percent return on assets.

Jay's disregard of financial constraints did not result from lack
of knowledge. Although he was not a trained accountant, he soon
learned the basics of analyzing financial statements, came up with
many of the company's creative financing schemes, and was skilled
in handling bankers. As Marcella said, "He could pick financial
statements apart just like that," and was always aware of Flying J's
financial status. Although he continually kept the company precar-
iously close to exceeding its liquidity, the controller rarely had to
tell him, "We can't do it. We're out of money." Even under those
circumstances, Jay could generally find a way out.

An Assessment of Jay's Early Skills and
Strategies as an Entrepreneur

An example that clearly demonstrates Jay's skill, ingenuity, and
tenacity in penetrating the West Coast retail gasoline market

Fastway station in Pendleton, Oregon, 1970.

comes from how he started the Fastway station in Pendleton, Oregon, under a joint ownership agreement with Craig, the younger brother next to him in age. Pendleton is a farming community located in the northeast portion of Oregon, then 45 miles or more from another town of any size. In 1970, the local gasoline retailers were branded dealers in the town's old section. They shared the market and marked up prices by nearly 50 percent. A gallon of gas was 38 cents in Pendleton versus 24 cents in Provo, Utah, where Craig attended college.

This was another opportunity Jay saw to grab a large share of a rural market with a cut-rate station, although it could not be self-service under Oregon law. An ideal location was found, a small 45-by-90-foot corner lot on the right-hand side of the highway going out of town. The house on the lot was dismantled in one day, after which Craig proceeded to put in the tanks for two diagonal islands that would front a trailer house serving as a pay station and live-in facility. As Craig was filling in sand around the tanks, the mayor, building inspector, and other city officials angrily gave him the order to stop construction. Likely at the behest of the local retail petroleum operators, they insisted that Craig immediately terminate work and leave town because Flying J did not have the proper building permits, did not meet the city's fire code, and violated other regulations. The city had an ordinance (presumably based on fire safety) mandating that filling stations be serviced by bulk-plant operators, not truck transports. This ordinance gave the branded wholesalers with local bulk plants control over all retail stations within city limits and, in essence, prohibited independent operators. The local officials were actually doing harm to their constituents. Lower Flying J fuel prices would reduce the local cost of living, a fact later acknowledged in the local newspaper.

Jay hired a lawyer who proved ineffective. Not to be denied, Jay personally researched the code and discovered that a bulk plant could sell retail to the public. Hence, he changed his building application to one for a bulk plant, received city approval, and proceeded with construction. Under the ordinance, the one obstacle facing Flying J was the requirement that a bulk plant be fenced. With limited space at the site, Jay's solution was to locate the bulk plant on a 12-foot strip at the back of the lot, fence it, and run pipes underground directly between the pumps and the bulk plant. Transports had difficulty dumping fuel in the bulk plant because of the confined space, but the station soon opened offering gas at 29.9 cents a gallon, 7 cents under the competition. As Craig stated, "We blew the fuses on every oil dealer in town." Within six months, Craig and Jay earned back their entire $25,000 investment, and the facility provided an extremely favorable cash flow for many years after.

From the beginning of his business ventures, Jay displayed another Call trait—that of being thrifty—and he continues to be adept at minimizing expenses. The tendency within the company is to work long hours and to be shorthanded. While on the road, especially in the early days, corporate officials worked from dawn to dusk and rarely got to bed before midnight. From the start and to this day employees travel in low-priced rental cars, sleep in budget motels, and fly economy class. On occasion in the early days, Jay would hitchhike from the nearest airport to one of his motels, and he continues to take care in obtaining favorable but fair terms in making rental or purchase agreements.

Although Jay initially skimped on operating expenses, he was more than generous in rewarding performance. Salaries were never high, often less than the industry average, but if the company did well, bonuses more than made up the difference. For the first eight years, 25 percent of each salaried employee's income was placed in a pension fund, a figure unheard of, especially in retail. Cash in the fund was used to purchase Flying J capital items that were then leased back to the company, causing the fund to grow at an enormous pace. This continued until pension laws passed in 1974 prohibited such generous practices. Now the company's pension plan is consistent with others in the industry.

Employees who helped build stations in the Northwest delight in telling stories about Jay—stories that reveal his affable nature, sense of humor, and total lack of self-importance. Once when he and several coworkers got the last room in a motel and someone had to sleep on the floor, they drew straws and Jay lost. As he laid out his blanket he remarked, "Here I am the CEO and I'm sleeping on the floor." On another occasion during a downpour, they drew straws to see who would go for pizza. He lost again, followed by a similar comment. He has been known to ride in the back of a pickup truck while his employees sat in the front, and he likes to spend time in Flying J lounges conversing with truck drivers. As John Lyddon, a millionaire friend, marveled, "His relationship with people extends from the bluest of collars to the top tenth of 1 percent of the people in this country, and yet he can be comfortable with all of them." Buzz adds, "Jay is as comfortable in a country western place as he is in the swankiest restaurant in town." Jay prefers to deal in small groups where he can interact one-on-one. Like other people, his behavior toward those he associates with is dictated largely by how he views them. He cares less about their wealth or position; his concern is more with their integrity.

Another primary factor underlying his early accomplishments was being able to recruit honest, reliable operators who generally stayed on the job for several years at a time. Retail outlets are typically at a disadvantage because most jobs start at or near minimum wage. Thus the applicants tend to be younger, inexperienced, untrained or unskilled, and/or those with poor work habits. They typically view their positions as short-term with limited financial rewards, resulting in minimal job commitment. A common corollary is that "shrinkage" through employee theft is a major problem.

Jay was able to offset some of these liabilities through using live-in operators, often married couples of retirement age, who wanted something to do or found they could not live on their retirement incomes. They were enticed by the free rent and the possibility of greatly exceeding the minimum wage since their incomes were based on gallons sold. As independent operators, not employees, they were not affected by wage and hour laws. Many couples would keep their businesses open 16 hours a day and be there during the night to guard the facilities. Theft or embezzlement was rare since the only monetary transaction between the owner and the dealer was payment for the number of gallons sold.

From his experience in Willard, Jay recognized that this was the optimum way to oversee and manage a facility. It was a formula crucial to his success in most of Flying J's first 50 stations. Then, in 1977, the federal government declared that such operators must be considered employees, the company was fined, and the practice stopped.

THE OPEC CRISIS

AND A

CHANGE IN DIRECTION

"You can't stand pat on a rising drawbridge."
—anonymous

*N*o climb to the top occurs without major hitches. One of the most challenging for Flying J resulted from the 1973 October Israeli-Arab war that eventually caused crude prices to quadruple and OPEC to shut off petroleum exports to the United States. It was perhaps the closest Jay ever came to losing his company.

In January of 1973, Buzz telephoned from California and said he was having trouble getting fuel. Since this had never been a problem in the past, Jay was in total disbelief. He immediately flew to Bakersfield and verified that fuel was scarce and independents were the first to be cut off. Shortages occurred more quickly on the West Coast than in Utah, and by October, nine Fastway stations in Buzz's region had to be closed. After the war broke out the same month, the shortages rapidly spread nationwide. Within two months, half of Flying J's 32 stations were boarded up. In just three months, Flying J gallons sold dropped from 3.2 million to 1.6 million. Fortunately, the shortage forced prices up so company sales and profits did not follow the same steep descent. Sales decreased by only 6.2 percent and profits by 3.2 percent.

The problem independents had in getting fuel was that branded refiners were obviously going to service their own retailers first, and without upstream refineries as a source of supply, independent companies were gradually being starved from lack of product. Recognizing that the independents were helpless, Congress passed a law in November setting up the Federal Energy Office to allocate petroleum during the crisis. This office issued regulations requiring refineries to distribute their output to their regular customers in the same ratio that existed during 1972. However, this never gave the intended help to the independents since their supply had already been greatly reduced by that date.

In what Teddy describes as "one of the most trying times for Jay," his first action was to bring Buzz back to the corporate office in October of 1973 and put him in charge of supply and transportation. In attempting to overcome the drastic shortages in California, Buzz had learned the ins and outs of wholesaling and refining, and trucking was an interest that dated from his youth. Maverik, Flying J's major wholesaler, had a surplus of diesel and residual fuel. (Residual fuel is the "heavy" bottom component in refining and is used primarily for heating, asphalt for roads, and other purposes.) The refinery manager offered Buzz more gasoline if he would take these other commodities (primarily diesel) off his hands. Flying J immediately purchased three more trucks, two for diesel and a special transport for hauling residuals, and the company began a vigorous effort in wholesaling. Buzz aggressively marketed diesel to industrial end users, initially construction firms engaged in road building and then to companies such as Portland Cement, Weyerhaeuser, Thiokol Chemical Corporation (the missile contractor 25 miles northwest of Brigham City, Utah), and the railroads.

As so often occurs when a disaster shuts down one market, opportunities open in others. Rather than being a temporary fix, Flying J's trucking and wholesaling became a permanent segment of its business. By 1979 the company had 17 trucks delivering fuel to wholesale customers and Flying J stations, and wholesale bulk sales produced $33 million in annual revenue.

Jay's next move to thwart the downside of the crisis was to persuade his banking allies to extend his credit lines. Then he had to find work for the 15 employees in his construction department. (The company was saving 20 to 25 percent on new facilities by being a licensed general contractor. The company designed most projects and supervised subcontractors who did most of the work.) Jay quickly put the architect-engineers and construction crew on projects other than fabricating stations. They began designing and building motels to place on undeveloped property attached to existing stations in Carson City and Reno, Nevada.

Fortunately, Jay was correct in assuming the gasoline shortage would be temporary, and supply started to recover in April of 1974. After adding only two stations in 1974 (Idaho Falls and Carson City), the company got back on track with four outlets in 1975, three in Oregon and one in Rock Springs, Wyoming.

By the end of 1975 Jay had 36 stations, a fleet of seven trucks, a 49-unit motel in Carson City, one with 72 units in Reno, three condominiums in Palm Springs, and additional property— mainly large apartment houses in Salt Lake City, Tacoma, Portland, and Corvallis. In seven years Jay's retail facilities nearly equaled Maverik's in number, and he was on safer ground by being more diversified in real estate. In recognition of this success, in July the company moved corporate Brigham City headquarters from a double-wide trailer house to a 5,150 square-foot modern structure. Ron, Buzz, and Paul were all made vice presidents, the accounting department was delighted with its first computer, and the company gave every evidence of being in business for the long haul.

Life in the Call Family

From 1968 to 1975, many changes occurred in Jay's home life. The family now lived in a fashionable hillside residence in Brigham City. Jay would spend Monday in the office, the family had dinner together that night with no friends over, and then he would be gone for the balance of the week. Initially on weekends, they traveled in a camper (later a motor home) to sites normally within a hundred miles of Brigham City. Most trips were to the southwest side of Bear Lake in Rich County, Utah, on waterfront property Teddy had found. Sometimes Jay would fly the family to Disneyland, their condo in Palm Springs, various parts of Mexico, or other locations for longer vacations. Yet, given a choice, the children always opted for trips in the motor home, preferably to their Bear Lake retreat.

When Thad and Crystal were children, adults frequently commented on their unusual courtesy and exemplary behavior. They had certain chores to do around the house and, knowing Jay, one of the most common was to help wash and wax the cars and airplanes. When Crystal was old enough to drive, Jay bought her a car with whitewall tires. She made the mistake of driving to see him when they were not clean, and received his anticipated response: "I am never going to get you another car with whitewall tires. Look at those things!" Likewise, when they visited their condo in Palm Springs the children knew that upon arrival their first activity was to clean inside and wash the sheets—no going out in the sun or visiting neighbors.

According to her offspring, Teddy "always put us first," gave them complete trust, allowed unusual independence, and early on treated them as adults. Both were loving parents and "fun to be around." Jay was the disciplinarian, but the children had such strong respect for him they rarely got out of line. They liked being around their dad because "he always loved life" and wanted them to water-ski, ride motorbikes, fly in his planes, and "try new things." He gave them the attitude that whatever you do in life, go for it and make it work. Teddy refers to him as "a good father

Jay beside one of his early fleet trucks.

Jay with Thad and Crystal.

when he was there, although he always had a lot on his mind."

Jay proved to be a role model for his children just as his father and uncle had been role models for him. Both children worked in the company while going to college and were constantly aware of Flying J's impressive rise and the enthusiasm Jay had for what he was doing. Crystal recalls how she was always proud to pass a Flying J station and considered the company the "best of the best." Looking back after receiving an MBA from the Harvard Business School, she stated, "Until I was 18, I thought entrepreneurship was all there was. I had no idea about what it would be like to work for IBM or AT&T." She and Thad later become successful in business both inside and outside the company.

From the beginning of their marriage, Teddy and Jay had an understanding that she would be in charge of their social life and manage the children while he looked after the company. They had many friends—business associates, neighbors, and others in the community. On weekends, they went dancing, rode motorcycles, and took short vacations. Besides Jay's long motorcycle trips to visit his operators throughout California and the Northwest, on occasion he, Buzz, and others (often with their wives) would motorcycle as far as Mexico's Baja peninsula. Buzz calls these trips "some of the highlights of my life."

With Jay being away, Teddy kept busy with sorority, school, and church activities. She also was a valuable business partner, a situation that likely dominated too much of their relationship. A downside of entrepreneurship is that the founder must work long hours and be out of town for days at a time. Long separations never bode well for a marriage, and after 13 years, differences started to develop, eventually resulting in divorce. The divorce sent chills throughout the company. For three or four months after, their founder and leader—widely acknowledged as the heart of the enterprise—rarely set foot on Flying J premises, spending most of his time in his Palm Desert condominium or in the Northwest, primarily Portland. Many started to wonder whether the malaise was permanent and the company would flounder.

Although Jay was in California or Oregon for most of the lat-

Phil, Buzz and Jay on a motorcycle trip to Cabo San Lucas, Mexico, spring 1989.

ter half of 1975, he kept in constant contact with Flying J's top officers by telephone. Eventually he returned to Brigham City, modified the company's double-wide trailer into comfortable living quarters, and settled back to handling business affairs, interrupted only by a severe case of hepatitis.

In 1976, the rapid expansion of the business resumed. A net of 10 stations was added, bringing the total to 46. Five of these were in northern California, purchased from Roy Brown (Teddy's brother) and John Wallace, the Soda Springs banker. The others were new units in California, Oregon, and Washington. During the year, the company broadened its scope of services by opening its first stand-alone lubrication center, a Fastway Lube in Salt Lake City. (They changed the name to J Lube in 1979 when all company facilities were switched to the Flying J flag.) Being one of the first to offer a facility specializing in no-appointment oil change and lubrication services, the company gave evidence that it was at the forefront of the industry.

In 1977, four stations were added, all first-class units with a contemporary design. Company sales rose 26 percent (exceeding

$50 million for the first time), and net profits advanced 70 percent, to just less than $700,000. Both the company and its leader were back on track and momentum was building.

Jay remarried in 1977. He considered his one major failure to be the divorce. Not wanting to repeat the experience, he was very cautious in dating. Through a longtime neighbor, Jim Stone, general manager of Thiokol (the solid-propellant missile company northwest of Brigham City), Jay was introduced to Tamra Compton, a public relations employee at the plant who

Jay and Tamra at their wedding.

eventually became his wife. Tamra rapidly gained the admiration of Jay's close colleagues and other acquaintances. Within a few weeks, they were referring to her as "a fireball and lots of fun" and "like Jay, a concerned, levelheaded unpretentious person." She is widely regarded as someone who is "easy to be around, sensitive, and both attractive and talented." Jay's friends acknowledge his good fortune, not only in business but also in selecting wives.

Changing the Company's Business Strategy

In the four years following the end of the 1973 fuel shortage, Flying J sales nearly tripled, going from $23.5 million to just over $65 million. Assets also tripled to more than $18 million. However, because long-term debt financed half of this increase, the company remained highly leveraged with debt amounting to more than twice owners' equity. Burdened by major interest payments and restricted by stagnant fuel prices due to price controls, Flying J's net income remained essentially flat. At the time, refineries were highly profitable, but retailers like Flying J were struggling, causing them to search for other means to increase their bottom line.

Jay, always on the lookout for new opportunities, quickly sensed the situation and aggressively moved the company into an area of investing that was extremely profitable at the time—real estate. As noted, company assets had tripled in four years while the number of stations increased less than 50 percent, going from 34 to 52. The balance of the asset gain was mostly in real property.

As Jay acknowledges, most of the bottom line on the company's financial statements during the mid and late 1970s came from buying and selling property. The volume of their real estate transactions varied from $3 million to $9 million a year through 1979. This activity accounted for only 10 percent of Flying J's sales but more than half of its profits. Since long-term capital gains on real estate were taxed at half the corporate earnings rate, brokering real estate became even more attractive. Jay did not view this new strategy as a permanent shift in company policy but as a useful hedge against the erratic oil market. In addition, real estate fit hand-in-glove with his petroleum interests because property acquisition is the first step in carrying forward retail expansion.

When building his initial stations, Jay discovered that in buying an appropriate sized half-acre lot on a busy intersection or highway, he was paying 90 percent of the asking price for the entire parcel, often consisting of several acres. Therefore, he began purchasing the entire piece and either developing the surplus (such as with the motels in Carson City and Reno) or selling it to other commercial developers. At the pace prices were increasing (especially in the late 1970s when inflation was 10 percent, later to climb to 22 percent), the company frequently sold the extra property surrounding a station for more than the price of the original parcel. Accordingly, beginning in 1970, Flying J always maintained a large property inventory, a situation that made its high debt-to-equity ratio not so ominous because it carried real estate on its books at the purchase price, not market value. Besides, the major disadvantage of real estate (lack of liquidity) was not a liability then because it was generally sellable within a few weeks. Because profitable real estate transactions held up the petroleum end of the business in the latter 1970s, Jay devoted much of his energy to this activity.

Initially Jay worked primarily with Bill Perry of Salt Lake City to purchase apartment houses in that community, Portland, Tacoma, and other cities. A variety of other property was bought and sold, mostly along the Wasatch Front. Initially the emphasis was on undeveloped residential property. Later they purchased office buildings and constructed industrial facilities at Decker Lake west of Salt Lake City. They gained approval for several large housing projects that were sold before construction started. In 1977, after Jay ended his ties with Bill, Flying J established a property division within the company and hired Ron Parker as director.

In 1978, a local project that excited the corporate staff was construction of an 84-unit motel with an adjoining restaurant at 715 West North Temple in Salt Lake City. The restaurant (leased to Perkins Cake & Steak) and the motel were on the property surrounding the original J Lube. To celebrate the completion, several days before the motel opened Flying J had a huge party for all employees, many of whom stayed in the rooms overnight.

Finding a New Direction

One of the most significant years in Flying J's history was 1979 when the Carter administration started to deregulate gasoline. Suddenly the retailing of fuel became attractive again. Fuel prices, responding to more normal supply and demand forces, jumped by 33 percent. With fuel more abundant, Flying J kept adding to its truck fleet, a move that boosted wholesale sales to $33 million, more than one-third of the company total. Overall sales also increased by one-third, and profits by 177 percent, exceeding $1 million for the first time.

This wholesaling and retailing boom plus profitable dealings in

Key Flying J employees reviewing the Ogden plaza design in late 1978.
Left to right: John Telford (head of construction), Marcella Hume (office manager),
Ron Brisendine (V.P. Retailing), Jay, Stan Weeks (controller), and
Buzz Germer (V.P. Supply and Transportation).

real estate were significant for Flying J, but meanwhile the year's two major developments affecting the company had little impact on its immediate bottom line. The first of these was almost as momentous for Flying J as when Jay quit Caribou Four Corners and started on his own. It involved the construction of a major truckstop in Ogden, an event that would forever change the company.

In one of his shrewd real estate moves, in the early 1970s Jay acquired just over 20 acres west of Ogden on 21st Street. Later, knowing that it would soon become the site of an interchange when Interstate 15 was extended north from Salt Lake City, he bought more surrounding property, bringing the total to 68 acres. Initially he intended to open a large trailer park and use the balance for industrial development. Buzz on several occasions had suggested they develop a truckstop, but Jay said the property was too valuable for that purpose and, being "Mr. Clean," he was reluctant to become involved in what was considered the unsavory end of fuel retailing.

However, by 1979 conditions had changed. The upward explosion of gasoline prices combined with fuel economy measures imposed by the federal government made it evident that national auto fuel consumption would not just level off but decline. The same would not occur with trucks because they were continuing to displace railroads in delivering freight. In addition, the I-15 extension was being completed, and no major truckstop was within 40 miles. Jay had given up the idea of a trailer park and was now considering a motel and possibly an industrial park. In giving consideration to a truckstop, it could be combined with a motel and a restaurant to consume 15 acres, leaving 50 more for industrial usage or to sell.

Buzz's concept was to enter the truckstop business just as they had begun with self-service stations. Keep prices low, invest to a minimum (unpaved driveways, trailers as pay stations, etc.), and offer only basic services. Once Jay became convinced that Ogden was an appropriate test market for a truckstop, he envisioned a more upscale facility, one that he would seek out if he were a driver. As Jay stated, "We built the stop having in mind dealing with people, not servicing trucks." As noted, conditions at the typical truckstop were deplorable: they were generally in the rundown, blighted areas of a community or highway; parking was limited; trucks competed with cars for space and service; facilities were dirty and showed lack of maintenance; bathrooms were small, lacked showers and offended the eyes and the nose; the food was generally marginal and served in grubby surroundings; and truckers were treated as the dredges of the traveling public.

Jay insisted that Flying J's truck plaza be just the opposite: generous parking (including overnight); service islands separating trucks from cars; tiled bathrooms and showers; attractive interior and exterior; a television lounge and rest area; ample telephones at strategic locations; and restaurant facilities at least the equal of national franchised chains. His intent was to make stops for truckers a pleasant experience where they could relax and feel at home rather than be constantly concerned about food contamination or safety. As one driver later told the Ogden plaza manager, "I love

View looking north at the Ogden, Utah, travel plaza.

View looking south at the Ogden, Utah, travel plaza.

to come here. It is one of the few places I go where I can take a shower and not have to worry about something crawling up my leg."

Many design features (later to be incorporated in Flying J's standard truckstop format) reflect facets of Jay's personality and values—attributes that significantly raised the standards in the truckstop industry. Above all, appearance and cleanliness backed by first-rate service are critical. Jay, always one to protect his own privacy, feels that the privacy of others should be respected. Hence, telephone booths are isolated so that conversations cannot be overheard, toilet stalls are totally enclosed, and private lounges are designed to allow truckers to relax with minimal interference.

Jay's concern was more than making truckers happy. Flying J architects designed the Ogden plaza not just for this clientele but also for local residents, tourists, and anyone else on the road. The goal was to integrate the company's real estate, motel, food, convenience store, and fuel retailing interests into a one-stop complex that offered quick service for the on-the-go traveler. Jay had high hopes for the new design, but little did he realize how this format would revolutionize the industry by converting minimal facilities designed solely for truckers into multipurpose, functional, tasteful plazas for the traveling public.

When completed, the Ogden design incorporated many features described above. Eight islands were equipped with 29 hoses (23 dispensing diesel) to avoid long lines at the pumps. A full-line C-store was constructed. A large recreation room was attached where truckers could unwind before getting back on the road. Bathroom facilities (including showers) were set aside exclusively for truckers based on Jay's cleanliness and privacy standards, and an 80-unit motel (then the equivalent of any in Ogden) was added with sizeable rooms at moderate prices. The stylish Tamarack restaurant and lounge seated 240 patrons and had a menu that contained a variety of well-prepared food. All facilities were of premium construction featuring brick veneer.

Such a plaza did not come without a hefty price tag—a nearly $3-million gamble representing over half of Flying J's net worth.

To minimize the risk and reduce enormous interest charges (then double-digit and on the rise), the company applied for and received a 15-year Weber County Industrial Revenue Bond for $1.8 million at 8.5 percent interest. Government-sponsored industrial revenue bonds were to provide low-interest, public-backed loans for businesses based on the assumption that the companies would create local jobs and generate significant sales and property tax revenues. In less than a year after go-ahead, on November 2, 1979, Flying J management and many of its employees glowed with pride at the plaza's ribbon cutting and open house. Jay, feeling that he had hit a commercial home run, nonchalantly told a reporter, "I just love to build things."

Although Flying J's management had wagered much of its future on the outcome of the new plaza, they were not just pleased but aghast at its immediate, almost instantaneous success. Gallons sold quickly jumped to 300,000 a month. The motel and restaurant became so popular with local residents and travelers that truckers comprised only a small portion of this trade. Jay was as content as a cat, realizing that he had discovered a market niche that would likely turn into gold. The feeling among company executives was much like a drilling crew hitting a gusher. The plaza served as the corporate flagship for the next several years and was a harbinger of things to come.

The Second Important Event of 1979

The second significant event in 1979 was also real estate–related. Through a series of fortuitous circumstances, Flying J acquired a choice 59-acre parcel in Pleasanton, California, at the junction of I-580 and I-680 in one of the most attractive regions surrounding San Francisco Bay. While visiting his sister in a community near Pleasanton, Ron Parker had at Jay's request looked for property in the area. After making several inquiries, Ron was informed by a broker that a prime parcel would soon come on the market since Boise Cascade Credit Corporation had recently taken it back when a builder defaulted. After seeing the property, Ron immediately

offered Boise Cascade $15,500 an acre or $900,000 with the provision that it be bought on contract with $100,000 down
(another major gamble since the purchase price was nearly one-
fifth of Flying J's total net worth). By telephone, Ron convinced
Jay to put up the earnest money. Jay, still in the hunt for attractive
real estate, considered it a prime motel location and began planning one with 200 units. The one drawback was that it had to be
rezoned, requiring several months of intensive effort. The wisdom
in making the transaction did not take long to become evident.
After the rezoning, with an outlay of only $200,000 in cash, the
company within days was receiving unsolicited offers in excess of
$4 million.

As stated in the December 1979 Flying J *Newsletter*, acquiring the Pleasanton property was "the most profitable single transaction Flying J had made." Although Jay and Ron did not then
realize it, picking up nearly $4 million in profits and cash flow
helped Flying J fund its next major venture, one that again
changed the direction of the company and propelled its growth.
Rather than adding a few stations at a time, Flying J, in one
action, acquired several corporations that boosted sales more than
200 percent in two years.

Important Management Lessons

Jay's decision to build a plaza with amenities pleasing to truckers
and the public at large was the result of basic lessons he had
learned through the years and are a key to understanding the
industry. As noted in chapter 2, because fuel products are difficult
to differentiate to the public, companies are restricted to attract-
ing customers primarily through price, location, and hospitality. In
the past, competition in the industry had been largely on price and
location. The success of the Ogden truckstop not only gave Flying
J an edge in the marketplace, but it forced the balance of the
industry to pay more attention to the third factor—hospitality.
The lesson, simple in logic but powerful in practice, is that if a
product is difficult to differentiate, differentiate the way in which

it is offered, primarily through the service provided.

In a fast-changing, technology-driven world, business leaders in the early 1980s and especially in the 1990s learned that the survival of their companies depended on their continuous ability to reinvent their organizations, product lines, and services. Jay displayed this flexibility in the 1970s when he refocused the direction of Flying J on several occasions. He is a prime example of the adage that the capacity to adapt is the sole means a company has to insure its longevity.

One enigma relating to Jay's thinking is whether it is mostly short term, where he is reacting to conditions and opportunities as they occur, or whether he develops long-term strategies that guide the company. When one reporter pressed him in 1980 regarding how his plans came about that led to getting into truckstops he responded, "I don't have anything figured except just to get up every day and try to find an opportunity." As we shall see, he was again being modest in reflecting on how he analyzes situations and on what he has accomplished.

Left to right: Hailey Maggelet, Tyson Call, Drew Maggelet, Alexis "Lexi" Maggelet, Tamra (Tamme) Call, Whitney Call, Erica Maggelet, and Jay.

Thad Call with his son Tyson by Jay's helicopter.

Chuck and Crystal Maggelet with their children: Alexis (left), twins Hailey and Lexis (behind held), and Drew (right).

Jay with the Maggelet twins in Montana.

Jay trout fishing in Montana.

Tamme and Jay vacationing in Europe.

Jay and Tamme in front of a dolled up car given by "friends."

Tamme, Sid Johnson, Jay, and Phil Adams when Jay was presented the car.

Tamme and Jay in Montana.

Tamme and Jay on his yacht off the West Coast of North America.

Fishing off Mexico in 1994.

Reuel Call (Jay's uncle) and Jay. (1994)

Jay with a king crab.

Jay inside one of his earlier helicopters.

Tamme and Jay, 1986.

Jay and Tamme on his motorcycle at their former home in Perry, Utah.

ACCELERATING GROWTH
THROUGH A
MAJOR ACQUISITION

"You can't make money consistently if you are uptight."
—Boone Pickens

$\mathcal{E}$ver since the 1973 oil crisis, Jay had felt vulnerable without a guaranteed fuel source for his company. He had watched the petroleum majors weather the turbulence in world oil markets and come out unscathed; as integrated, almost self-contained suppliers, the majors had their own wells, refineries, and retail outlets. Jay yearned to get Flying J in this position, waiting for the time when assets could be added upstream (refineries and wells) to provide the needed protection. Any such action would be a major step for a small retailer because exploration and refining require massive initial investments that provide delayed returns compared to quick cash flows in fuel retailing.

In the late fall of 1979, Buzz Germer and Ron Brisendine were in Spokane looking for truckstop property. While having

lunch with a Flying J fuel supplier, they learned that the company
he represented would likely be coming up for sale. The firm, a
composite of four corporations engaged in all aspects of petroleum and natural gas production and retailing, was known as
Thunderbird Resources. It was headquartered in Denver and
owned by Inter-City Gas Ltd. (ICG), a Canadian corporation that
was Canada's largest propane distributor. ICG was dissatisfied
with its U.S. operations because of low profitability and increasingly stringent requirements imposed by the Environmental
Protection Agency. In addition, its U.S. holdings consisted of an
odd mixture of businesses brought about by three separate mergers from 1969 through 1974, and ICG intended to concentrate
on propane. Although 14 percent of ICG's corporate assets were
tied up in its U.S. divisions—generating 28 percent of company
1978 revenues—they only accounted for 3 percent of total operating profits. Even worse, ICG's U.S. refineries were losing money
in 1979 at a time when small refineries (such as those of Maverik)
were extremely profitable. This profitability was largely due to the
federal allotment program favoring small refineries, a program
that existed during the period of petroleum shortages and price
controls in the latter half of the 1970s.

Ron and Buzz, knowing that Thunderbird had nearly twice
the sales and more than three times the assets of Flying J, assumed
a takeover was highly unlikely. However, knowing that Jay would
not reject a long shot if it opened the door to becoming an integrated oil company, they quickly made him aware of the pending
sale. Jay was immediately impressed, especially because
Thunderbird's operations fit into the commercial part of the
country he knew best. Accordingly, Jay quickly called
Thunderbird offices in Denver. After officials there denied any
knowledge of the matter, he phoned ICG headquarters in
Winnipeg, Canada, and they invited him up to discuss a buyout of
their U.S. operations involving assets of between $30 and $40
million.

Thunderbird's fixed assets consisted of a 5,500 barrel-a-day
(b/d) refinery in Cut Bank, Montana, a 4,500 b/d refinery in

Williston, North Dakota, a closed refinery in Kevin, Montana, a 65,000 gallon-per-day natural gas processing plant at Cut Bank, 32 propane marketing outlets, and 53 other marketing facilities, primarily bulk plants and small service stations. All were in Montana, North Dakota, Idaho, Washington, and Oregon except for a lube oil blending plant in Louisiana. In addition, Thunderbird operated a major wholesale business through its jobber network of 300 stations. As a potential bonus, Flying J would gain part ownership in more than 60 oil and gas wells and obtain lease and drilling rights in major oil-producing fields in Montana and North Dakota surrounding the two operating refineries. Thunderbird's disparate facilities accumulated through the various mergers included several nonpetroleum businesses such as auto parts outlets in Montana and North Dakota, plywood mills, and a uranium mine. The latter two were not part of the sale.

After analyzing Thunderbird's holdings, Jay was convinced that the individual pieces were worth more than the whole based on the asking price. The gasoline stations in North Dakota and Montana (under such names as Thunderbird, Vista, Westco, Westland, and Domestic) were mostly small and dated with varying formats. However, those in the Northwest, though aging, were generally first class and on valuable property, often of generous proportions. The stations, wholesale business, and propane bulk plants were important, but the prizes Flying J coveted were the refineries and oil and gas fields.

For any company to take over another exceeding twice its size requires courage and aplomb, but Flying J's financial condition made it even more of a David and Goliath situation. Obtaining major new loans and credit lines would be difficult because Flying J owners' equity was only $5.145 million or one-sixth of the asking price. The company's long-term debt was already $8.8 million, resulting in an unfavorable debt-to-equity ratio of 1.72 to 1 and making any added borrowing difficult. The Thunderbird acquisition would boost this ratio to six dollars in long-term debt to every dollar in equity, a ratio generally deemed death threatening even when interest rates are low, and in 1980, they were well over 12 percent and approaching the highest level in decades. Furthermore, Thunderbird's refineries were losing approximately $1 million per month, losses that if continued would swallow up Flying J's assets within a year.

To get ICG to sign, Jay calculated he needed $10 million in hand. He hoped to obtain half from First Security Bank, at this point his major source of credit. After presenting a proposal to First Security officials, Jay, Buzz, and Paul met with George Eccles, the conservative company president, and on the spot, he agreed to a $5 million loan with one proviso: Flying J must get Thunderbird's current banker, First Bank of Minneapolis, to finance the other half by granting a term loan and line of credit. Jay realized this would be difficult because First Bank was already concerned over Thunderbird's growing refinery losses and Flying J's questionable balance sheet. As much as the bank's loan officers admired Flying J's management team, they felt the company was too weak financially to take on such an acquisition. Following several visits with First Bank officers remaining noncommittal, Jay, Buzz, and Paul realized they were at a dead end and would have to find other options.

While flying back to Brigham City after receiving First Bank's final rejection, Jay looked at his dejected companions and said, "Let's not go home. Let's go to Winnipeg, lay our cards on the table, and see if we can negotiate a deal." Knowing they had nothing to lose, both quickly agreed, and in midair Jay directed the pilot to turn north for what they realized was their last opportunity to salvage an agreement.

Their only hope was for ICG to carry more of the credit, a move that would likely keep First Security Bank in with their $5 million. After skillful negotiation, Jay got ICG representatives to finance $17,777,000, more than half of the $31,561,000 purchase price. The debt owed ICG was to be paid in installments over ten years with one balloon $6 million payment for the Williston refinery to be made on December 31, 1980. With such favorable financing, not only did First Security stay in but Jay also got First Bank to come back on board. The change in ownership

was set to occur on April 1, 1980. Paul Brown, Flying J's controller and Jay's right-hand man in the negotiations, insisted that
interest on the ICG loan be at fixed rather than variable rates and
that an indemnity clause be inserted relieving Flying J of any
responsibility for prior violations of EPA standards. With ICG
eager to sell, the negotiations closed within 24 hours.

The two features Paul had incorporated in the agreement
proved to be life jackets for Flying J that would keep it from
drowning in its own red ink. The 13.5 percent fixed interest rate
proved a savior when prime rates jumped to 22 percent in 1981.
Later, under the indemnity clause, ICG was forced to pay two-
thirds of the refineries' hazardous waste cleanup costs when the
EPA found all three sites in violation, a development that proved
costly to both companies.

Raising Funds and Making Thunderbird Profitable

Buoyed on for several weeks by the excitement of entering the big
time and becoming a significant full-line petroleum company, Jay,
Buzz, and Paul now realized they faced the formidable task of
turning around a much larger, diverse enterprise that was poorly
integrated and ineffectively managed. Under the terms of the
agreement, one of the major financial challenges was to arrange
for the $6 million payment at the end of December for the
Williston refinery. Again, Jay resorted to local government help,
getting Williams County executives in North Dakota to issue
industrial revenue bonds. Although this arrangement did not
reduce company debt, it kept it in good standing with ICG, and it
provided lower interest rates for this portion of Flying J's total
obligation.

Still, Flying J's risks in the acquisition were more than financial. The company to date was primarily a retailer with no experience in exploration, drilling, and refining. Jay's assumption was
that with the help of consultants, Buzz's quick learning skills, and
current Thunderbird employees, Flying J could handle the situation. However, he underestimated Thunderbird's weak manage-

ment, especially in refining, where they even lacked an engineer to handle technical operations. Thus, it turned out Flying J was both understaffed and lacking in talent to deal with its immediate problems. As Buzz recalled, "In retrospect, we did not know anything about the operations of a company that size."

What nearly proved fatal for Flying J was miscalculating the required working capital. Coming from a retailing background, Flying J had always been flush with cash resulting from being able to sell delivered fuel before the supplier was paid. With exploration, production, and refining, the cash flow situation is just the opposite. The time between incurring costs and receiving cash is often measured in weeks, depending on how long fuel is kept in inventory. On occasion, when a refinery is shaky financially, crude suppliers will demand a guaranteed line of credit or payment on delivery to avoid losses if a refiner defaults.

Without considering the bank loans, lines of credit, and need for increased working capital, the amount owed ICG alone was nearly three-and-a-half-times Flying J's book value. Jay, knowing it would be tight financially, had several cards to play. First, he knew his situation was better than it appeared because Flying J assets were significantly understated, especially the Pleasanton property, resulting from real estate being carried on company books at the purchase price, not current market value. To reduce the debt quickly, Jay sent Ron Parker to California and told him not to return until the Pleasanton site was sold. By selling Pleasanton, the Reno and Carson City motels, and other real estate, $8.1 million was raised, a sum still far from adequate. However, these quick actions and results did gain Jay considerable credence with the banks and others involved in financing the takeover. The amount obtained was consistent with his earlier estimates, confirming that he had not inflated the company's financial position.

Besides raising more working capital, Flying J quickly had to take two other steps to stay solvent. One was to sell as many of ICG's properties as possible while retaining the refineries and oil and gas properties. The other was to stop the financial hemorrhaging, especially that coming from the refineries. ICG's inadequate cost accounting made it difficult to know where to start. Under ICG's system, all company retail outlets had to absorb refinery products at cost rather than at market price. Since the refineries were in a loss position, this resulted in fuel costs that exceeded local wholesale prices. Thus, refining losses were passing on to their retail units, making it essentially impossible to pin down the profitability of any particular facility. Thunderbird's records were in such disarray that when Jay brought Ron Parker up to inventory ICG properties, he had trouble not only in finding many deeds but also in locating the real estate involved. After losing precious time, Flying J was able to identify weaker units and put them up for sale. Other more marketable assets—several prime retail stations, the lube plant in Louisiana, and the auto parts outlets—were put on the selling block and soon picked up by others.

Although Flying J's market timing of the acquisition could hardly have been worse, Jay was well aware of the risks. When the takeover occurred, he admitted that "5,000 barrel-a-day refineries are considered poor performers in today's market." As noted, government support for small refineries was being withdrawn, and crude oil prices were on the rise. The exploding crude prices were especially disheartening. From 1979 through 1981, crude jumped from $13 to $34 a barrel on the world market following the overthrow of the Shah in Iran, the Iran-Iraqi war, and the fuel shortages in the U.S. brought on primarily by hoarding. In addition, the Carter administration had started to deregulate U.S. fuel prices. The effect of crude prices tripling threatened to push Flying J working capital beyond anything the company could muster.

Cutting the refinery losses became the company's top priority. How to gain efficiency and improve marketing were the primary problems. The refineries were small and lacked sophistication, even after ICG had recently spent $10 million in upgrades to produce unleaded gasoline. Their major drawback was a limited capacity to produce high-value fuel products, resulting in large quantities of residual fuel that had limited demand. Without the

capacity to recycle residuals to obtain additional diesel and gasoline, the primary option left for Buzz was to sell the residuals to
large industrial plants for heating. Even this alternative soon vanished when rising crude rates forced residual prices to exceed the
alternative—natural gas. Thus, Flying J was left trying to unload
its surplus residuals to refineries that could convert them to products more in demand.

Due to Flying J's lack of experience, Thunderbird's weak
management, and limited funds to make much-needed capital
improvements, the situation appeared hopeless. As an added problem, just before the acquisition, the Cut Bank refinery was on
strike and Flying J had no experience dealing with labor unions.
Having gone this far, Jay, Buzz, and other Flying J personnel
became more resolute. With the aid of a consultant, they immediately began to reduce inventories, change pricing, improve
accounting controls, strengthen marketing, and incorporate Jay's
management philosophy, all of which contributed to plugging up
the $1 million monthly drain. As Buzz summarized, "Eventually
we went from tremendous losses every month to some profits."
Unfortunately, the national conditions for small refiners continued
to deteriorate, dashing any prospects for long-term profitability.

In a situation reminiscent of what Jay had experienced seven
years earlier, Flying J's bright future became dimmed by changes
in the U.S. economy and world events beyond company control.
As deregulation progressed, for the first nine months after the
Thunderbird transaction only 20 percent of the crude fed to
Flying J refineries was subject to price and allocation controls.
Then, on January 1, 1981, total deregulation took place. This
occurred when the national demand for motor fuel was decreasing
for the first time in U.S. history except during the two world wars.
The drop-off in demand was due to higher fuel prices, improvements in engine efficiency, 55 miles-per-hour speed limits, and
occasional gasoline shortages. The result was a national decrease
of 8.8 percent in motor vehicle fuel consumption from 1979
through 1982.

This damage to small petroleum companies was further inten-

Thunderbird refinery at Cut Bank, Montana.

sified by the economic conditions plaguing the country as a whole. The economic depression in the three years following 1980 was the most severe since before World War II. It was driven by high interest rates, runaway inflation (nearly 30 percent in three years), peak postwar unemployment, and a slump in retail sales. The windfall profits tax passed by Congress hurt the majors, but their large oil inventories and reserves had been turned into black gold by price deregulation. For the small refiners, the depression added one more nail to their coffins because few owned enough wells to be self-sufficient, and only the most efficient, sophisticated, and well-located refineries could survive. One-third of all U.S. refineries closed between 1982 and 1984, most of the smaller variety.

Flying J's struggle to keep its refineries open soon ended, although Jay was proud of the progress they had made in becoming "one of the most innovative refiners in the Intermountain area." The company shut down Williston in January of 1983 and Cut Bank two months later. Buzz resisted phasing them out, calling the closing of Williston a "real heartbreaker." "One of the hardest things I ever had to do was layoff a hundred or so people in the middle of winter" in a small town where it would "devastate the economy." "But," he added, "I realized that the sun would still come up, my family still liked me, and life would go on." The intent was eventually to reopen the refineries, but conditions never improved and the company later dismantled them to prevent their being community eyesores. The natural gas plant at Cut Bank (doing business primarily for Montana Power) continued to operate until 1987.

As of this writing, the company still has the original Thunderbird oil and gas properties within the production and exploration division. Also, the wholesale fuel business acquired from Thunderbird (especially that in the Northwest) continues to grow. Many retail fuel outlets obtained in the buyout contributed to Flying J profits into the 1990s, but few remain as viable properties. The propane division was profitable but never found a comfortable niche in corporate strategy and was sold in two separate transactions in 1983 when Flying J experienced another frightening money crunch.

The human resources strain on the company resulting from absorbing Thunderbird was more than just managing the refineries. The number of employees required to handle the combined operations placed such demands on the corporate office that many of the new hires and transfers were housed in three temporary trailers for more than a year until better space was located. Initially a dozen or so of Thunderbird's management personnel relocated to offices in Brigham City. However, combining the systems and work forces of the two companies took its toll and most left within a few years. Those that stayed attributed it to Jay who, as one stated, "is hard to beat as a boss."

The merger placed extreme pressure on all aspects of Flying J's administration, especially those units that were marginal at the time. In some divisions, such as accounting and electronic data processing, Thunderbird—being a larger, more diverse company—had more comprehensive, sophisticated systems. This forced Flying J to upgrade equipment, adopt new systems, and adjust to a fourfold increase in workload.

The pros and cons of taking over Thunderbird have been debated in the company ever since. With few exceptions Flying J participants agree that the major benefit came from gaining experience in refining, production, and exploration, thus positioning the company to make important acquisitions in the future. Without question, the biggest drawback was a delay in the timetable to expand the truckstop and travel plaza business. Like most growth companies getting into new ventures, acquiring Thunderbird was a broadening experience, valuable in terms of learning, but a short-run sacrifice in profits, given other opportunities at the time. Nevertheless, as Jay stated, "It put the company on the petroleum map." In effect, it forced Flying J into a sink-or-swim position where the choices were to go defunct or to transform a small company into a big one and thus pass over a difficult hurdle all businesses must cross if they are to continue to rise in their industries. Of course, this maturing did not occur without considerable strife. It absorbed most of the company attention for three years, much longer than anticipated. However, as Bob Heiner, then executive vice president of First Security Bank, stated, "Thunderbird kicked off some of the bigger things to come for Flying J."

Surprisingly, even with the strain of the merger, Flying J made a net annual after-tax profit for the next few years, an accomplishment deserving high praise. By selling off many of Thunderbird's assets and improving the efficiency of the remaining business units, Flying J was able to handle its debt, insurmountable as it had appeared. In 1980, the year in which the merger occurred, company sales jumped by 130 percent, going from $88.6 million to nearly $204 million. Sales rose by another 34 percent in 1981, reaching $273 million. Profits, a meager half of a percent or less in 1980 and 1981, were slim but rewarding, although the situation kept company executives and their creditors uneasy. Cash became especially critical when national interest rates reached 20 percent and fuel suppliers cut the payment cycle from 30 to 10 days or even less, thus eliminating the financial windfall Jay had used since the 1960s to expand his growing enterprise.

The Aftermath of Thunderbird

The problems the company experienced with Thunderbird failed to put a clamp on Jay's aggressive management style. He never stopped expanding current operations or investing in new ventures. To handle the growing work force, a corporate annex with executive offices and space for 105 employees was completed in 1982. In 1983, the Salt Lake motel was expanded by 39 units, and an 84-unit motel was built the next year at a recently acquired Boise truckstop. A waterslide entertainment center north of the Ogden travel plaza proved a failure, but another sideline business, rental of video equipment and movies, paid big dividends for several years.

As is common with entrepreneurial ventures, the opportunity to enter the video business came along almost by chance. After hearing of video successes in California, in 1982 the company introduced the "Flying J Video Club" in several of its Montana stores. These were the first rental video outlets in that area. Each store, starting with just 100 titles, enjoyed brisk business that surprised Flying J executives. Within three years, the company had video outlets at 35 Flying J locations and all of its motels plus 121 dealerships spread out over 11 states. The company's video inventory soon exceeded 37,000 tapes with nearly 2,000 VCR rental machines. By 1985, Flying J was one of the top ten national video rental chains, verifying that the business was extremely profitable. Then, as often occurs with a product or business that is on the perimeter of a company's primary interest, Flying J did not keep up with the intense competition created by soaring demand. In several years, sales at supermarkets and stores devoted solely to video rental took the lead. However, Flying J kept its video operations in the black until the early 1990s before backing out in 1993. Thad, Jay's son, worked in the video division for several years before leaving to manage a travel plaza.

The company also undertook a major program to update its self-service gasoline stations. Many facilities still contained apartments for the operators even though the government prohibited this arrangement in 1977. Strapped for cash, Flying J was slow to convert this space to the convenience store format that began to dominate automobile gasoline retailing, led by such companies as 7-Eleven and Circle K. With corporate sales dropping 15 percent in 1982, Flying J started an extensive remodeling program late in the year. Within three years, 28 of Flying J's 55 C-stores were refurbished and up to the caliber of the competition.

As alluded to earlier, a combination of the continual struggle in digesting Thunderbird plus adverse national economic conditions (especially for petroleum) forced the company in 1983 to go through a 10 percent reduction in force. Flying J's employment in 1980 had jumped from 350 to more than 700 when Thunderbird workers were acquired. Even though many were released as facilities were sold or the refineries closed, Flying J still had a labor surplus. Partially due to reduced national demand, gallons sold at company outlets dropped 21 percent from 1982 to 1983 and sales by 10 percent (although travel plaza sales were up 50 percent). Jay realized that for the first time in company history he had to dismiss part of his work force and reduce payroll costs. Besides cutting employment, the company froze wages and salaries and eliminated bonuses. Given Jay's caring nature and concern for employees, these decisions caused him considerable grief. As one employee stated, "The layoffs were a very emotional thing for Jay." It made him resolve to keep employment in check, never wanting such an occurrence to happen again.

Keeping the Focus on Truckstops and Travel Plazas

Burdensome as the Thunderbird acquisition was, Jay never discarded his plans to be a leader in truckstops and travel plazas. Almost before the ink dried on the purchase agreement, Jay again proved willing to ignore financial constraints if he spotted an opportunity he knew would be a treasure trove—this time the acquisition of another truck plaza. In December of 1980, a large truckstop in Boise came up for sale. Located on a desirable 26-acre site just off the I-84 freeway overlooking the city, it suffered from being poorly managed and in need of remodeling. Being unprofitable, it had changed owners twice in just a few months. With all his personal assets and those of the company tied up in Thunderbird, Jay realized it would be impossible to finance the purchase. After evaluating several options, he offered half owner-ship to his friend Sterling Jardine of Jardine Petroleum based on the provision that Flying J would be the operator. Once Sterling agreed, Jay was able to come up with the balance of the obligation and the purchase was made. To turn the business around, the two partners immediately made $700,000 in improvements and, within a year, sales more than doubled followed by a 27 percent climb the year after. In July of 1983, Flying J bought Sterling's share, and the Boise plaza became one of Flying J's most valuable assets. (In 1998, Flying J sold the Boise plaza and opened a larger one in Caldwell, Idaho, 25 miles west on I-84.)

In part, buying Thunderbird aided Jay's plan to focus on truckstops since two were included in the acquisition, one at Great Falls, Montana, and the other at Williston, North Dakota. Soon after, Flying J purchased two more, the first in Beach, North Dakota, in 1983; and, one year later, the second on 21st South west of I-15 in Salt Lake City. In addition, a "mini" truckstop was constructed in Orland, California, in 1983. However, the most important truckstop development during this period was to take place at Post Falls, Idaho.

The Post Falls travel plaza was significant because it replaced the one in Ogden as the company's format of the future. Buzz and Ron located the 7.5-acre lot just across the Washington border in northern Idaho, 30 miles from Spokane. The company searched for property near the state line because Washington motor fuel taxes were much higher than Idaho's, and truckers would fill up before entering the state. The Post Falls design set a new standard for travel plaza decor and was developed with the intent of maxi-mizing convenience, comfort, and hospitality. It differed from Ogden in that the restaurant, C-store, and cashier station were all under one roof, reducing both construction costs and required staffing. The restaurant had seating for 112 and an expanded C-store with 1,000 feet of shelf space. Gasoline customers had access to 10 pumps and diesel customers to 24. Parking was considered ample (later to prove marginal) with space for 80 rigs.

The criterion used to evaluate every aspect of the design was to minimize labor costs. In such an elaborate complex, keeping fuel prices cut-rate meant labor savings had to be achieved. Accordingly, at Post Falls all customers—including those at the restaurant—pay bills at one front desk location. Jay, still somewhat unsure of how truckstops would take hold nationally, was intent on keeping payroll costs down if demand turned out to be less than projected. Much to his relief, soon after the $2 million facility opened in October of 1982, the manager had to hire 70 employees to service the growing trade.

In all, Post Falls generated another retail marvel that rivaled Ogden in significance. Business flourished and the company started to share in Jay's belief that some day Flying J would have a travel plaza network covering the entire nation.

BUILDING A
MANAGEMENT TEAM

*"There's no limit to what you can accomplish if you
don't care who gets the credit."*
—Robert Townsend,
former Avis CEO and author of
Further Up the Organization

The period between 1980 and 1984 not only tested Flying J's
financial strength, but also seriously challenged the depth and
capability of its management. Not long after acquiring
Thunderbird, Flying J started to experience the strains of becoming a large enterprise. Small, entrepreneurial, rapid-growth companies generally become stalled in their expansion unless many of
the original management team are replaced or choose to go in
other directions. The skills involved in running a small business
are rarely those that can be converted to managing a complex
bureaucracy. At a certain point in a company's growth, the sheer
size of its operations forces a need for more staff specialists and
managers with a broad perspective of how to push the organization forward and compete with established, more asset-rich com-

panies. Coordinating diverse operations, overseeing the work of others, and hammering out decisions in a committee setting are the anathemas of the independent, hands-on, commercial adventurer who avoids restraints in bringing pet projects to the marketplace. Studies show that the typical start-up will go through 2.7 executive officers before going public, largely due to the changing nature of management at each stage of a company's growth.

Jay's search for additional executive talent began in earnest after garnering Thunderbird. He knew he was at his best when identifying opportunities, making deals, initiating projects, and receiving quick feedback through "being able to see my tracks." In addition, he still acknowledges his dislike for paperwork, his reluctance to be tied down by appointments and meetings, and his boredom with the coordination and maintenance demands of complex organizations. At the time of the Thunderbird buyout, he told a reporter, "I am not one to stagnate or just run the shop." He clearly created the vision and provided the leadership that brought the company to this level, but he was also aware that for his own peace of mind he must find someone to handle day-to-day operations. Later, when executives from Thunderbird failed to fill key company roles, the urgency in expanding and upgrading his management team intensified.

A related reason for start-up failure is that the founders are too often unwilling to diminish their own roles or share power with others. As his wife Tamra stated, "Jay realized a long time ago that the company wouldn't grow very big if he did that." An early trait Jay displayed (one that both his father and Uncle Reuel lacked) was to avoid being a one-man show. Jay, always open to new talent, once told Sterling Jardine, his original partner in the Boise plaza, "Don't be afraid to hire people smarter than you are. A lot of managers can't do that because they have to feel they're the smartest." Jay will not let pride or the need for power stand in the way of letting others make decisions, an attribute rarely found in zealous, expansive-minded entrepreneurs. He is a prime example of the saying "A managers hire A-plus subordinates and B managers hire C subordinates."

Jay's philosophy on the selection, supervision, and growth of subordinates is the key to understanding how he built Flying J's management structure and determined his own company role. These concepts, a by-product of his personality, are principles that guided Flying J's staffing from the beginning through his tenure as company president.

Hire Potential, Not Experienced Executives Demanding High Pay

At the time of the Thunderbird merger, a statement reflecting Jay's philosophy appeared in the May 1980 company newsletter:

> Some managers don't want experienced people working for them. Experienced people will often tell you why something can't be done. The person without experience often doesn't know the reasons why something can't be done and goes ahead and finds a way to do it.

Jay was obviously influenced by problems he had experienced with high-profile executives such as several CPAs hired to run the financial side of the company and by Buzz's amazing development. Jay was also disturbed by the failure of Thunderbird executives to fit in. Thunderbird carry-overs and various financial advisors were constantly urging Jay to run the company in ways more conventional by establishing detailed controls, reducing risks, and building up a cadre of professional specialists. Jay, however, was convinced that, given Flying J's restrictive financial resources, he was better off hiring individuals with potential but limited experience—novices who were quick to learn through the challenges he offered. He refused to create a revolving door through paying high salaries for experienced managers who soon left because of their difficulty in adjusting to Flying J's work ethic and corporate culture. Perceptive, young, new employees learned to accept their initial low pay and, by taking advantage of the wide-open opportunities to assume responsibility, ending up far ahead in the long run. Jay's attitude is reflected in a common management adage: "When you can't afford to hire the best, hire the young who are going to be the best."

An example of how Jay opened the door to those with potential (similar to the Buzz Germer story) involves John Scales, current president of Flying J Oil and Gas Inc. John, armed with a master's degree in geology and two years of work experience, was hired by Andy Morgan, who was then running the former Thunderbird oil and gas properties for Flying J. Andy had been a consultant to Thunderbird, and Jay hired him to manage the properties once Flying J obtained ownership. Andy, 63 at the time, decided one year later to retire. Then, as John explained,

> Jay flew up one day, came to my office, and told me Andy was thinking of retiring. "We need to replace him," Jay commented. "Do you know of anyone to recommend?" My response was that I would like to give it some thought. (To a certain degree, that was Jay's way of approaching me.) Then he said, "Well, if you're given the job, what would you do?"
>
> It caught me by surprise, but I listed six or seven things that would improve our operations right away. Jay smiled and said, "That sounds good to me. Let's go tell everybody that you're now the General Manager and VP." So, we went in and told everyone, most of whom were as shocked as I was. As soon as the announcement was made, I drove Jay to the airport and he departed with little comment other than the usual goodbyes.
>
> In the next few weeks, I was engaged in implementing the things we had talked about. I never heard from Jay and I started to get worried. As time went on, I became more concerned and decided I just had to call. When I did, Jay asked

what I wanted. After I explained that I was concerned because we hadn't spoken for six weeks or so, he responded, "I thought if you needed help or had a question, you'd get a hold of me." Our relationship has never changed to this day.

Motivate by Perks Such as Bonuses and Favorable Retirement Programs

Unable (and to some degree unwilling) to pay high salaries, Jay looked to systems that reward employees for outstanding performance and keep them committed. He knew from experience that no incentive could exceed that of being an owner or part owner of a company. In addition, he felt that those employees who help develop a business should share in the booty as the company grows. Thus, beginning in the early 1970s, Jay established several programs to achieve these results.

The first was a bonus plan set up to provide cash awards each December. When the company was small, Jay would delight those in attendance by passing out the bonuses to the wives of the male employees at the annual Christmas party. The unique feature of the bonus plan is that it included all employees, even those paid hourly. When the program was being designed, Jay emphasized that workers should be rewarded for what they do, not solely for longevity. Under the system, management annually determined the amount to be distributed, and allocations were made to each facility or organizational unit. Bonuses varied based on a point system that took into consideration the work unit's productivity, individual performance, longevity, and so forth. The amount received was on a par with that of other industrial firms, but other programs (especially those of retailers) rarely include all employees. For some hourly workers the bonus could add up to a month or more of their annual incomes, a significant incentive since approximately 80 percent of Flying J's work force is paid hourly.

Another enticing incentive, dating back to the founding of the corporation, was Flying J's stock ownership program. As

noted earlier, initially Flying J managers received 25 percent annually in a private retirement system designed with the help of a benefits consulting firm. This changed in 1974, when Congress
passed the Employee Retirement Income Security Act (ERISA)
that forced the restructuring of many private pension programs
and set up specific rules defining types of acceptable retirement
plans. Flying J adopted one of these, an Employee Stock
Ownership Plan (ESOP). The defined contribution portion of this
ESOP was financed entirely by the company, which made an
annual contribution of 3 percent of all wages and salaries. This distribution was in cash or Flying J common stock at the discretion
of the board of directors. To date all contributions have been in
the form of common stock.

This was supplemented by a profit-sharing contribution (again
in stock) approved annually by the board of directors based on
company profits for the year. The amount varied annually, generally ranging from 2 to 3 percent of compensation, making the total
company retirement contribution 5 to 6 percent. New employees
served one year to qualify for the program and became fully vested
(having total ownership of their portions) after seven years of service. In February of 2001, the company froze the ESOP program
and replaced it with a deferred payment bonus program that offers
similar rewards although not all employees are included.

In addition, Flying J has a 401(k) plan based on federal
guidelines wherein employees can put up to 12 percent of their
pretax pay in a personal savings program that likewise is intended
for retirement. The dollars placed into the plan are generally tax
deferred until withdrawn. For those who participate, Flying J adds
50 cents to each dollar a person contributes up to an annual maximum company contribution of $1,200. Flying J's current 401(k)
contributions exceed $800,000 annually.

Be the Most Efficient by Doing More with Less

Numerous examples exist of Flying J's "leanness." In the
Thunderbird takeover, Flying J's negotiating team was outnum-

bered four to one by the bankers, ICG officials, lawyers, and others representing the opposition. After Flying J started operating the refineries and other properties, former Thunderbird employees could not understand how they could do it with so few workers. Years later in several major negotiations, the opposition financial advisors and lawyers kept wondering when Flying J's small team would be supplemented by additional experts. When visiting the former Brigham City corporate office for the first time, executives from other companies were generally awestruck when they found such a modest facility and limited staffing; Flying J's former head of data processing recalled "large companies, such as AT&T and IBM coming in here and going away shaking their heads in amazement at what we do." The former head of Flying J's construction division was once quizzed by a disbelieving professional, "You're doing $75 million in construction and you have 13 people to do it?"

Doing more with less has characterized the company since Jay opened his first station. Employees are expected to take responsibility and do whatever it takes to get the job done. Managers take pride in keeping the company lean. As Richard Peterson, vice president for fuel marketing, supply, and distribution, stated, "It would break my heart if we got big and inefficient like other large companies in the industry."

This frugal approach obviously has some risks. It likely contributed to the ease two employees had in embezzling more than $600,000 from the company between 1990 and 1992. However, keeping trim has enabled Flying J to attain the enviable record of being the low-cost provider even though it has some of the most elaborate, expensive travel plazas on the interstate system.

Growing Talent through Empowering Others

Long before empowering others became popular in management literature, Jay was carrying the concept to extremes far beyond what most managers practice today. It is a theme Jay continually pounds away at with his staff and disbelieving executives from other companies. As noted in the examples of Buzz Germer and John Scales, once Jay gives subordinates an assignment he provides them almost total freedom to act on their own. No greater harm can be inflicted on an organization than to have employees afraid to act because of their own perceived limitations or because their supervisors might come down on them. One reason Jay so willingly delegates authority is that he finds great satisfaction in seeing those around him grow and take on responsibility. As he has often stated, "If they succeed, I will succeed." Under Jay's leadership, Flying J created the ideal climate for skill development, one where being aggressive is encouraged, performance is rewarded, and employees are trusted and held on a loose rein. Phil Adams, current company president, summarized his feelings and those of many employees:

> In a nutshell, I think the success of Flying J is the environment that Jay created. I know a lot of others here feel the same way. I am not that motivated by money, but I am awfully motivated by having an environment to do what I want to do and having the freedom to do it. There's nothing I dream of or want to do that I can't accomplish in this environment.

Similar statements came from Richard Peterson: "There's as much opportunity here as any company in the world. I really feel that way. You can take as much responsibility as you want."

Both Phil and Jay acknowledge that Flying J has made its share of mistakes by turning over freedom to the wrong people, but the losses involved are minuscule when compared to the gains through unlocking the creativity and initiative of the work force.

Lead by Example and Learn from Mistakes

Jay's relationship with subordinates is unique. Following his usual practice, rather than directing them in their work, he lets them learn from their mistakes. On one occasion while flying back to Brigham City after visiting several stations with a district manager, Jay commented to him that he needed to do a better job in keeping the facilities in first-class condition. The concerned manager immediately wanted to know what Jay had in mind. Rather than going into detail, Jay responded, "If I need to do that I've got the wrong man for the job."

More than once Jay has allowed a subordinate to pick a bad site or make a poor investment to see if the individual could later recognize the error and correct it. On other occasions, he has been known to get on his knees and clean tar off a carpet or empty overflowing garbage cans at a C-store just to let the managers know they are not meeting his cleanliness standards. One of the often-repeated stories was an occasion when a manager was called down to the station in the middle of the night by an assistant manager. Much to his alarm, the manager found Jay on his knees scrubbing the floor. When he entered, Jay looked at him and remarked, "This floor is terrible. It's filthy dirty. If you want to keep your job, you'd better keep it clean. I don't want customers seeing this." Although Jay can be very subtle, his associates claim they have never seen him angry. He disciplines by first warning the person several times. If no improvement is made, Jay becomes candid and to the point. With a lean staff, those who eventually fail to pull their own weight are advised to seek other employment.

The employees' great respect for Jay is due in part to his personal values, leadership, and the way he treats them. Crystal observes, "One thing I admire about Dad is that he has always done what is best for the employees." As Buzz states, "The work force has the utmost respect for him and would step in front of a bus if necessary." Diana Hansen, a district manager over several C-stores, commented (referring to the company's colors), "I would bleed orange if someone cut me." Ron Parker, vice president of real estate, explained it another way: "I have so much respect for him in the way he treats people that if he asked me to be on top of Mt. Timpanogos to make a deal on New Year's Eve, I'd find a way to be there."

Realigning his Management Team

In 1982, Jay's top management team consisted of five vice presidents: Buzz, the self-made expert in refining, transportation, and production, was vice president of supply and production; Ron Brissendine continued in his position as vice president of retailing; Paul Brown, a CPA in charge of financial affairs, was executive vice president; Ron Parker, the person who made the Pleasanton real estate deal, was vice president of real estate; and Jack Dailey, an experienced manager from the outside, was vice president of the propane division. Ron Brissendine, Paul, and Buzz had all been vice presidents since 1975 and were the trio Jay relied on to help run the company.

Each trio member was less inclined to take risks than Jay. In this respect, their conservatism served as valuable checks on his occasional exuberance. However, it gave him concern over who

Flying J officers, 1981. Left to right: Jack Dailey (V.P. Propane Division), Ron Brisendine (V.P. Retailing), lower front Paul Brown (Executive Vice President), Buzz Germer (V.P. Supply and Transportation), Jay, and Ron Parker (V.P. Real Estate)

should be next in line for president, especially if the strategy of becoming the leader in retail travel plazas was to be pursued. Buzz was comfortable with his niche in the nonretail upstream side of the business, and besides, this segment of the company led in sales and would become even more important in the future. Ron had been essential in developing the retail end of Flying J, but he had not prepared himself for more responsibility, especially the growing scope envisioned by Jay. In 1984, Ron left, partially due to his desire to go on his own. Within a short time, he had constructed two C-stores in North Dakota and started a video rental business. This left Paul, who was the apparent choice after being promoted to executive vice president.

Paul originally joined Flying J in January 1973 when the company had only 30 cut-rate stations. His background included an accounting degree from the University of Utah, six years of experience with the Peat Marwick public accounting firm, and 18 months as a controller for a newspaper. He was Flying J's first full-time CPA. He conducted Flying J's initial audit, put in a variety of cost controls, and got the financial end of the company on a more professional footing. In 1976, he left to start his own accounting firm in Brigham City. He came back to Flying J as senior vice president just prior to the Thunderbird acquisition. Demonstrating the confidence Jay had in Paul, he made him executive vice president two years later.

Everyone in the company has high admiration for Paul. During the time he was outside the company, several different CPAs took over the controllership, but (except for Phil Adams) all had difficulty handling the job and soon left, often not of their own choosing. In the early years, Paul provided a professionalism that was important in dealing with bank officers, other lending institutions, suppliers, and public accounting firms. Inside the company, he was respected for his hard work, honesty, and broad knowledge of financial matters. He is not one to showboat or be pretentious, and his fellow managers always appreciated the stability he provided during his various tenures.

Any company contemplating a major expansion, such as that

Flying J had in mind, faces two critical problems: obtaining the capital to make the enormous investments, and having the knowledge to use these funds effectively in the marketplace. Paul clearly had the qualifications to handle the financial side, but Jay had some reservations regarding his background and interest in being the driving force behind expanding the retail end of the business.

A Newcomer Rises to the Top

At this critical juncture, Jay was still looking for someone to step forward who shared his broad vision and had the potential to take on the prodigious task of challenging the large, asset-rich oil companies. Few expected this void to be filled by an accountant, still in his late twenties, who arrived on the scene during the Thunderbird acquisition. J Phillip (Phil) Adams graduated from Brigham City's Box Elder High School in 1973. Five years later, he received an accounting degree from Utah State University. Then he became employed in Paul Brown's Brigham City CPA firm. Within a year, Paul returned to Flying J and took his young protégé with him. In July of 1981, when Paul moved up to be executive vice president, Phil was made chief financial officer. One year later, after showing remarkable knowledge and poise for a newcomer, Phil was made vice president and treasurer.

Being aggressive, full of confidence, and business-wise for his age, Phil took full advantage of the open environment Jay had created. Paul pushed Phil ahead when he demonstrated unusual skill in handling financial matters, and Jay took immediate notice of this youngster who had a good grasp of where the company was financially, sought responsibility, and gave evidence of being able to handle company-wide problems.

A coworker recalled an incident that demonstrated Phil's early leadership potential a year or two after his employment. In a session with the top management of a major public accounting firm, Phil not only held his own but also drew the partners over to Flying J's position. Jay had been disappointed with the accounting firm's preliminary audit that would put the company in the position of paying much higher corporate income taxes than had been anticipated. Being cash poor from the Thunderbird takeover, every dollar saved in taxes was crucial. The coworker and Phil prepared several schedules and arranged to meet with a partner of the firm and his assistants in Salt Lake City. As the coworker related,

> At the start of the meeting, Phil managed to take charge. He explained what we wanted, provided the backup schedules, and made clear that this is where we wanted to be. I was absolutely dumbfounded because when Phil got done they agreed. I was in utter amazement that a kid of that age and limited experience could approach professionals of that stature and win.

As Phil continued to exhibit skill in handing vital company issues, Jay started to see in him the take-charge executive he was seeking. Knowing that Phil had to broaden his background before he would be prepared for a company-wide executive position, Jay started giving him assignments in other departments. However, when sales declined and debt increased in 1983, Jay took the surprising step of passing over Ron Brissendine and making Phil vice president for retailing. Few CEOs would have thrust a young accountant into such a position, but Jay saw this as the opportunity to give Phil the required seasoning and the chance to demonstrate the capacity, interest, and drive necessary to become head of the company. Jay hoped that Phil would succeed because he was the one who most fervently shared his belief that Flying J could dominate the national truckstop and travel plaza business. In addition, Phil, like Jay, displayed a willingness to gamble against what most viewed as high odds if the outcomes were sufficiently attractive.

Letting a young manager leapfrog over higher-ups had its repercussions. Jay made the move against the advice of many of his associates. Ron Brissendine left the company, and Paul Brown departed soon afterwards in November 1984, to return to school and pursue an MBA degree. Jay, Phil, and others implored him to stay, but he was resolute in making the change, hoping to advance his career through more schooling. After Paul left, for nearly a decade Flying J experienced repeated turnover in the controller and chief financial officer positions, causing chronic turmoil in this branch of the company. Fortunately, in 1993 Phil talked Paul into returning as chief financial officer, thus restoring the stability and professionalism required.

Immediately after Phil became head of retailing, Jay worked with him on a daily basis, much closer than he had any prior employee. He was intent on giving Phil the training he needed and seeing how committed he was to making Flying J a premier company. During the next eight years as Phil continued to mature, Jay gradually withdrew from day-to-day operations knowing he had made the right decision regarding his successor. Phil was appointed as executive vice president in 1987 and as president in 1991. Jay continues as chairman of the board. Once a week he attends an executive council meeting with top company managers, but counts on Phil to run the business. Jay has total confidence in his company president. Jay once stated that he did not think there were ten managers in the country that could have done for Flying J what Phil has accomplished.

Phil is broadly recognized for his brilliance, aggressiveness, ability to "look at the big picture ten years down the road," and willingness to make hard decisions regardless of personal conse-quences. As one close business associate stated, "I do not know anyone sharper and smarter or anyone who is able to deal with so many issues." He is constantly searching to find better ways to serve customers and, in the process, expand company services. Like Jay, he would rather learn from mistakes than be unwilling to take action. Executives of other companies consider Phil articu-late, a superb negotiator, and extremely capable in representing Flying J interests. Insiders view him as determined and demand-ing, qualities that keep the company on its toes and provide the internal drive for Flying J to stay ahead of competitors. Early in his career, Phil was criticized for how he handled relations with some

Jay Call,
President and CEO (1987)

J. Phillip Adams,
Executive Vice President (1987)

John R. Scales,
Vice President and General
Manager Flying J Exploration &
Production, Inc. (1987)

R.E. (Buzz) Germer,
Senior Vice President and
President of Big West Oil Company
(1987)

fellow employees. His aggressiveness occasionally creates friction, but over the years, he has won universal respect for his leadership in creating travel plazas unrivaled in service and quality and for making the company number one in North America for diesel sales.

Phil turned into an extremely impressive corporate executive while still less than 40, due in part to the environment that Jay provided and the truckstop market niche awaiting development. He has unlimited aspirations regarding what Flying J will become in the future. Not long after he became retailing vice president when Flying J owners' equity was less than $30 million, he was driving with several other employees past a building owned by one of the giant petroleum companies. In looking up at the structure he commented, "We're going to be that big some day." The others chuckled before they discerned that Phil was dead serious.

Other Young Talent Brought on Board in the Early 1980s

In this transition period, several other young managers became employed who would also fill vital roles. The first company full-time lawyer was Barre Burgon, hired in 1980 just after the merger with Thunderbird. Barre had been practicing law for a year or two when the outside attorney who was then handling Flying J legal matters did not want to go full time with the company and recommended Barre for the position. Once hired, Barre was immediately thrown into a variety of legal problems involving real estate, acquisitions, Department of Energy regulations, employment law, taxation, and other matters. He developed a broad grasp of corporate legal affairs and has been an extremely valuable member of Flying J's top management team, currently serving as vice president, corporate counsel, and secretary and treasurer. His skills became particularly important in the decade beginning in 1984 when the company faced numerous legal challenges. Barre is universally respected in the company for his exceptional congeniality, integrity, concern for others, and being a team player. Outsiders consider him a strong strategist and one who can be tough in negotiations.

In 1981, Richard Petersen, an accountant with several years of experience, came to work for Flying J in the propane division. After the division was sold, he became head of retail accounting at the time when it was decentralized to be closer to operations. In the next decade, as Flying J facilities became dispersed throughout the country, retail accounting became more significant, especially since each large plaza requires sophisticated systems to keep on top of thousands of daily financial transactions. During this decade, Flying J has led the retail fuel and C-store industry in the development of real-time, automated systems, due in part to the insight and leadership that Richard provides. Today he holds the position of vice president for fuel marketing, supply, and distribution.

At the same time, several other young managers helped strengthen Flying J operations, such as John Scales in oil and gas operations, Ron Parker in real estate, and Don Rognon who at various times was credit manager, operated the motels, and served as an assistant to Jay. More key figures would sign on in the next few years who added diversity and quality to Flying J's management, preparing the company to reach billion-dollar sales.

Back on Track after the Downturn

In 1984 all aspects of Flying J operations improved, moving sales up 13.3 percent. At the start of 1985, the company's retail business consisted of 65 gasoline stations, 28 convenience stores, 7 travel plazas, 7 restaurants, 4 motels, and a major video business. However, retail sales were only 42 percent of total corporate sales with the majority coming from wholesale operations, the refinery, and the oil and gas fields. As would be the case for several years, this upstream side of the business provided the sales and profits enabling the retail segments to expand. The importance of the upstream operations was such that there is reason to question whether retail would have made it on its own.

Building long-term debt continued to be the force behind expansion. In 1984 long-term debt was the highest in Flying J's history at just over $40 million. On the bright side, the status of owners' equity had vastly improved. Book value had increased nearly three times from 1980 when Thunderbird was acquired, and was now a more reassuring $16 million. Thus, the debt-to-equity ratio dropped to a less threatening 2.5 to 1. Even with the sell-off or closure of a significant portion of Thunderbird facilities (especially during 1983), Flying J's fixed assets at the beginning of 1985 were just slightly lower than immediately after the merger. The company felt more secure, but the problem remained of how to gain the financing to becoming a big-time player in retailing.

THE HUSKY TAKEOVER

"True leaders are those who can make the best out of adversity."

—anonymous

With high debt and limited equity, Jay's plans to develop an interstate network of plazas seemed farfetched at best. With each plaza then costing a minimum of $2.5 million, the company would require a massive influx of capital before it could build more than a half dozen a year, a pace far too slow to assume the leadership position envisioned. Thus, company executives were again on the lookout for ways to accelerate growth, knowing it would be difficult if not impossible to get sufficient backing from investment bankers or lending institutions. Given Flying J's experience with Thunderbird, many questioned whether making another acquisition was the way to go. However, those knowing Jay recognized that if the right deal came along the trauma of Thunderbird would be forgotten. Both Phil and Jay realized that the only way to obtain the needed capital to exploit their travel plaza plans was to take drastic steps to multiply Flying J's assets, even if it meant again putting everything on the line.

Uncovering an Opportunity

In the spring of 1984, Jay was vacationing with Tamra in Palm Springs, California, when he read in the *Wall Street Journal* that Husky Ltd. of Canada had just struck a deal with Marathon Oil Company (part of U.S. Steel, since renamed USX), to buy Husky's U.S. oil and gas production properties. Jay had been anticipating such a sale since 1979 when Husky Ltd. was acquired in a hostile takeover by another Canadian firm (Nova Ltd.). He expected Nova soon to tire of their U.S. operations and put them on the selling block. Husky's U.S. properties were spread throughout 13 western states, but its three refineries were close by in Utah and Wyoming. Knowing that Marathon had turned down buying Husky's refineries and that these would be difficult to sell, Jay sensed a deal in the making, much like Thunderbird. After informing Tamra, he immediately told her to start packing. He thought the remaining Husky properties might be obtained at a bargain, and he was going to be on the company doorstep with an offer.

After gaining more information, within a few days Jay contacted Husky officials in Denver. They let him know that the RMT[1] division containing their U.S. refineries and retail properties was being prepared for sale, and the price would be around $100 million. The high price tag alarmed Jay, even though he knew it was inflated. Still it was far outside the range of what Flying J could finance. Besides, he was somewhat incensed because when making the inquiry their reaction was to be insulted at such a small fry approaching them. Jay had given them little business in the past, and they were apparently unaware of his reputation. Accordingly, he temporarily dropped the matter.

Husky's U.S. Holdings

The similarities between Thunderbird and RMT are striking: both had their operations in the western states, an area where Flying J

---

1. Abbreviation for refining, marketing, and transportation.

had a presence and market expertise; both were owned by Canadian firms eager to dispose of their U.S. holdings; both had truckstops, service stations, refineries, and a wholesale distribution network; and both had let their facilities run down in anticipation of their disposal.

The big differences between Thunderbird and RMT were in size and quality. Each had three refineries, but Husky's were more than three times the size, and two were still operating in the depressed market of 1984; Husky's retail network was larger (400 branded dealers), had wider name recognition, and was more strategically dispersed in 13 western states. However, of the retail outlets, Husky owned only 24 of the truckstops and 12 service stations. Nevertheless, Husky's truck plazas comprised the largest single western U.S. truckstop chain next to Unocal Corporation, and many were at prime locations in California, Washington, Colorado, and Wyoming. With RMT's three refineries and a respected truckstop business, Flying J's pursuit eventually became as enthusiastic as a wolf stalking a much larger wounded prey.

Of Husky's three refineries, the one with a 12,000 b/d capacity in Cody, Wyoming, was shut down in the small refinery slaughter of 1982. The one in Cheyenne, Wyoming—large for the Rocky Mountain area at 35,000 b/d—was expensive to operate and needed upgrading, both deterrents to Flying J. The third refinery, situated in North Salt Lake City, was half the size of the one in Cheyenne, but it could easily be upgraded to 25,000 b/d. Under normal conditions, the Cheyenne refinery would command a high price, but it attracted little interest in the negative refinery market at the time. Without question, the one property that Flying J prized, viewing it as the missing weapon in its battle with regional competitors, was the North Salt Lake refinery.

Although the North Salt Lake refinery began operation in 1948, it was still one of the newest on the Wasatch Front. It had been upgraded several times, especially a major addition in 1962. Three other features made it particularly attractive.

First, getting petroleum products into the state from outside Utah was unusually difficult because only one small incoming

Flying J refinery in North Salt Lake.

pipeline existed and that was already full. Transportation by truck is more expensive and thus not cost effective when long distances are involved. In comparison with Salt Lake City, Denver has several incoming pipelines from sources primarily in the Gulf region where large, highly efficient refineries are plentiful, keeping fuel prices low. With supply more limited in Utah, refineries experience higher profit margins.

The second factor resulted from a situation similar to that which Flying J experienced with Thunderbird. By grouping all refining costs into one common account, Husky made it difficult to determine the profitability of each facility. The first thing Buzz did after gaining access to Husky's books was to segregate costs by facility or profit center. In the process, he learned that the North Salt Lake refinery was far more profitable than the other two.

The North Salt Lake refinery's final advantage was that, through its being located in Utah, Jay would finally attain the guaranteed source of supply he had sought for over a decade. This protection was especially important for his stations in Utah and Idaho. (An outgoing pipeline from Salt Lake City provides petroleum products to several locations in Idaho. Idaho has no refineries and thus depends on Utah for approximately 70 percent of its fuel products.) As an added bonus, the refinery was not only profitable, but it had the potential of doing much better.

Husky began as a small family petroleum business under Glenn Nielsen of Cody, Wyoming. His business roots had some parallels to Reuel Call's. Nielsen started with small Wyoming refineries before expanding into surrounding states and Canada. Simultaneously, he became a major petroleum retailer in both regions, eventually turning the company into a major public corporation. The corporation's most valuable properties turned out to be Canadian oil and gas leases. In 1978, Nova staged an unfriendly takeover at a time when all petroleum companies were struggling, making it possible for Nova to obtain Husky for the bargain basement price of well under $1 billion. Like Inter-City Gas, Nova wanted to concentrate in Canada and was eager to exit the U.S. because of environmental protection laws, low prices, and intense competition.

The Bumps and Starts in Making a Deal

Within a year after Jay had approached Husky, a report appeared in newspapers and trade journals stating that John Grambling had purchased RMT. However, within a month the deal was nullified when bankers discovered that Grambling had misrepresented his financing and could not make good on the agreement. Embarrassed by what appeared to be a swindle, Husky now looked for a buyer with known integrity. The bad publicity, continued losses by RMT, and slowness in disposing of the company scared off most suitors resulting in Husky offering Flying J a reduced price in hopes of a quick agreement. Of course, Jay was pleased, but he knew an uphill struggle remained to get financing.

Recognizing his cash and collateral dilemma, Jay attempted to purchase only the truckstops, but Husky would not bite. Then he added the North Salt Lake refinery to the offer with the hope he could get financing for both. Again, Husky refused. Intent on disposing of all U.S. operations, they were not going to be caught again with the dredges left to sell.

Not be denied, Jay displayed his usual daring by making a nonrefundable down payment of $2 million to seal the transaction, and on June 5, 1985, Flying J and Husky signed a purchase contract for all RMT holdings with a closing date of August 1. Flying J executives were elated, but they knew the more difficult task remained of getting financing. Initially they went back to their major banking sources and then those of RMT's, but Flying J's debt-heavy balance sheet and the fact that more than 100 U.S. refineries had been closed during the past four years made them reluctant. Flying J's primary banker, First Security Corporation, gave some promises, but would not finance more than it had with Thunderbird. Several other bankers showed interest, but on each occasion, they withdrew before the settlement date. In a situation reminiscent of Thunderbird, Jay was forced to ask Husky for an extension so he could regroup and look for other possible capital.

At the time, Husky was becoming more desperate for someone to run RMT. Several top managers had resigned during the John Grambling fiasco, losses were accumulating, and their U.S. operations were in disarray. As an immediate out, Husky invited proposals from potential owners and internal managers to show what they would do if selected to become operators. All recommendations except Flying J's were to build up RMT by investing more capital, adding personnel, and modernizing facilities. Jay and his staff proposed just the opposite, in part because they recognized RMT had to slim down before they had a chance to have their offer accepted. Their plan called for disposing of weak units, reducing inventories, and speeding up cash flow.

Not only did Husky like Flying J's proposals, but after interacting with Flying J's management for several months, Husky now had full trust in Jay and the other officers. At this point, Husky was looking for operators with integrity as much as those with financial strength. Accordingly, on September 1, 1985, Flying J began managing RMT on a fee basis assuming that a switch in ownership would shortly occur. Jay was made president with Buzz and Phil as vice presidents.

Wanting to reduce its obligations and also ease Flying J's financial burdens, Husky urged Jay to proceed immediately with the downsizing necessary to shrink required working capital, now nearly $100 million in inventories and accounts receivable. Flying J's job was made difficult because Husky added the condition that employees could not be terminated even though the company was significantly overstaffed. Nevertheless, within three months the nearly $100 million needed to support operations was greatly cut back. With this decrease, Flying J officials finally felt they had RMT in their pocket. In fact, Jay was in Salt Lake City about to take the podium to inform Husky truckstop managers of his plans when he received word that the Bank of New England, a key financial institution in the negotiations, had withdrawn and the deal was off.

The gloom in the Flying J camp lasted only briefly before they renewed their search for other ways to decrease the required capital. Options they considered were to lessen their obligations by $5 million through locating a buyer for the Cheyenne refinery and to get a crude supplier to defer taking payments. Initially they could not find a buyer for the refinery, but the major crude supplier came in line by agreeing not to bill RMT until revenue was earned from sale of the refined output. Even with such a favorable arrangement, Flying J still came up short.

As the end of the calendar year approached, some Flying J officials began to lose hope. Husky officials were also becoming more frantic, apparently because they faced a major tax liability (associated with the Marathon sale two years earlier) if the company continued operating in the U.S. after December 31. Accordingly, like Thunderbird, at the last minute Husky stepped in (much to the delight of Flying J executives) and guaranteed a revolving credit line at a major bank with the agreement that Flying J would take over this commitment within 18 months. This action brought all parties back to the table, resulting in a New Year's Eve transfer of ownership that propelled Flying J into being the largest independent oil company in the Mountain West, an accomplishment that surprised everyone and had Flying J officials celebrating.

The final sale figures were nearly identical to those of the Thunderbird purchase except for the $39 million in revolving credit lines at two banks. The price for RMT was $31,969,000, within a half million of the Thunderbird agreement. The portion carried by the seller was $16,255,000 or $1.5 million less than with Thunderbird. The major difference—the larger bank credit guarantees—was made necessary by Husky's more extensive operations, mandating mammoth inventories to support the refineries.

To make the purchase, Flying J again absorbed enormous debt. Of the $32,969,000 purchase price, Flying J paid approximately $11.5 million in cash, leaving nearly $21.5 million in debt plus the $39 million in letters of credit with two banks. Fortunately, the debt to Husky was long term with payments starting in 1988 and extending through 1992.

The final negotiations were led by Phil and Barre Burgon with Jay in attendance. They faced 14 principals and lawyers representing Husky's interests. A president of First Security Bank (Jay's close friend) called and warned him that Flying J could not possibly do well with his young vice presidents competing against the high-powered Husky team that included several New York lawyers. Jay told his friend just to relax and see what happens. Then, according to Jay, "We just came out great. We didn't have one single problem with the deal, not one."

The $70 million agreement ($31 million purchase price and $39 million in guaranteed credit lines) again represented a Herculean challenge for a company with a net worth of one-fourth this total. The banks, fearful of the situation, required that both Flying J and its primary owner put up the company's entire stock and all their combined assets (including Jay's home) as loan guarantees. Furthermore, the company and Jay were restricted from obtaining other loans, making major capital acquisitions, or undertaking significant business expansions until the indebtedness was worked down. Flying J and its CEO were bound in financial straitjackets until nearly $20 million in debt and loan guarantees could be eliminated. For someone who might lose everything gained through more than 25 years of hard work, Jay remained amazingly calm. Tamra noticed that he experienced unusual pressure during the long drawn out process of acquiring Husky, but "he didn't stew a lot, he was just glad when it was finally put to bed."

Again, the financial shackles the banks placed on Jay did not stop him from making other purchases, especially if one involved an airplane. He was planning to become partners with a friend to purchase a Citation jet for $1.15 million when the friend could not obtain financing and backed out. Jay reluctantly approached his bank and, as he stated, "They went through the roof." He then called another friend in California who had the resources to be a co-owner, but he turned him down, although he asked Jay how much he needed to make the deal. When Jay's response was $1 million, the friend asked Jay for his bank account number and said he would wire the amount. Having trust in Jay, his friend told him there was no need for collateral or to sign a note. He could just send the money back in a couple of months, which Jay did.

As ominous as the figures appear, Jay and the company initially assumed less risk with Husky than with Thunderbird. The appraised value of RMT's 24 truckstops ($28.5 million), 12 stations and three land parcels ($4.7 million), three refineries ($7.85 million), pipeline distribution system connecting Cheyenne to various parts of Nebraska ($5 million), terminal in Boise ($465,000), and other properties topped out at nearly $50 million versus the $20 million fixed asset figure used in arriving at the purchase agreement. Furthermore, this was $19 million more than the entire purchase price, and it turned out that many of these properties were undervalued. Hence, by just selling the pieces, Flying J had a possible $30 million cushion if Jay chose to liquidate.

Aware of the financial crunch that would immediately follow the takeover, Flying J was able to tie down buyers for the two most valuable RMT assets: the Cheyenne refinery and the distribution pipeline. Both exchanged hands on a common closing date of March 1, 1986. A group led by a top executive of Tosco Corporation bought the Cheyenne refinery for $5 million, and the distribution pipeline was dealt to Continental Pipeline Company (Conoco) for approximately the same amount. For Flying J to rid itself of the burden of supporting these operations was almost as important as their sales. By disposing of other Husky properties (including a Denver station for $1 million), within a few months Flying J had $13.3 million to help cover the most pressing financial obligations stipulated in the purchase agreement.

When word of the acquisition reached the streets, Jay's acquaintances were again dumbfounded. One associate outside the company declared, "My gawd, how can Jay even think about doing that?" As Ed Swapp, one of Flying J's employees, stated, "When you have to put even your home on the line, most multi-millionaires would say, 'Hey, wait a minute,' but not Jay." Fred Greener, Flying J's primary accountant involved in the details of meshing the two companies, stated, "Husky did guarantee many of the loans, but it was still an ant eating an elephant." Gaining Husky properties gave Jay, Phil, and others the elation of feeling they had finally built company coffers sufficient to pursue aggressively their grandiose travel plaza strategy.

Facing Another Crisis

Unfortunately, as with Thunderbird, the timing of the acquisition could not have been worse. Just after the company's New Year's Eve victory celebration the world price of crude dropped as rapidly as it had risen in 1980, again threatening the company's solvency. Oil on the world market was overabundant and overpriced. As noted in chapter 2, Saudi Arabia, holding one-third of the free world's reserves, had been hurt most by OPEC's efforts to restrict production. When the Saudis decided to let prices fluctuate on the open world market, crude dropped from $28 a barrel in January to $12 on March 31, 1986, one of the swiftest declines ever. For companies with large inventories that had not been hedged (agreements to buy or sell crude at a future specified price), it was a disaster of the greatest proportions. For Flying J, desperate to cut Husky's operating losses and to sell properties to ease its financial strain, nothing could have been more damaging or untimely.

Soon after the purchase agreement was signed and while inventories were being downsized, RMT still had more than a million barrels, mainly of crude oil, in storage. With crude dropping $16 a barrel and gasoline prices plummeting 36 percent, Flying J had inventory paper losses of $20 million within three months of the takeover, nearly two-thirds of the amount paid for the entire RMT division. The fall was partially offset by the $13.3 million gained through sale of the properties mentioned earlier, but these dollars were required to meet loan guarantees. To help alleviate the situation, Flying J quickly sold its Salt Lake City motel, prime real estate at a Bountiful, Utah, I-15 exit, and many former RMT assets. During the year resulting from the value of fuel inventories plummeting, facilities being sold and written off, and debt payments accelerating, Flying J's assets decreased by $30.5 million.

As many Flying J employees recall, it was the most hectic period in the company's history. Buzz, Flying J's transitional manager who was temporarily located in Denver, Colorado, stated, "I have never been so busy in my life." As Barre Burgon, Flying J's corporate attorney, described it, "We were running from one end of the continent to the other and stopping in between to get things sold, bought, and borrowed." The hectic stop-and-go negotiations followed by the most chaotic petroleum market in at least six years put Flying J's management to its strongest test.

Fortunately, Jay and his colleagues were equal to the challenge. Actions by Buzz helped save the company as it teetered on the edge of falling into financial insolvency. When Flying J sold the Cheyenne refinery to Tosco, the buyers put $1 million down. Based on processing agreements, Flying J was to receive the balance in refined products. The new owners were most eager to dispose of the bottoms, primarily asphalt. Buzz, with a keen sense of market trends (asphalt demand had been low and the price weak but he felt it would turn around), arranged a three-year agreement to purchase asphalt at a price tied to crude oil. Asphalt prices on the open market rose almost as rapidly as crude prices fell, and within three years the company recouped more than $10 million of its losses.

All in all, Flying J was much better prepared to turn around Husky's operations than it had been Thunderbird's. Buzz now had a firm foundation in refining and the entire corporate office, having gone through one merger, was better set to handle another. Furthermore, the company avoided the overstaffing that accompanied the absorption of Thunderbird. Flying J boosted the productivity at all acquired facilities even though only a small fraction of RMT's employees were retained.

Assessing the Results

Evaluating the initial financial impact of acquiring RMT is awkward because of the timing of Flying J's fiscal year, which ends on January 31. With the exchange in ownership occurring on December 31, 1985, only one month (January) of Flying J's 1986 financial statements includes data involving Husky. Accordingly, all of RMT's assets and liabilities were reflected in Flying J's 1986 financial statements just prior to when many of the properties were sold and debts reduced. In addition, the spring of 1986 was when petroleum companies saw their profits tumble due to crude prices dropping 50 percent, and national gasoline consumption continued on the downswing, decreasing in Utah by 10 percent between 1984 and 1986. Still, combined company statements for fiscal years 1986 and 1987 show that during the two years, Flying J sales rose more than $100 million ($236 to $339 million). However, only $10 million or 10 percent of the gain came from retail sales even though the number of truckstops increased from 7 to 33. The large boost in nonretail sales came from the North Salt Lake refinery, wholesale operations, and Buzz's adroitness in selling asphalt.

The 1986 drop in crude prices caused major petroleum companies to experience the lowest combined annual profit gain in their histories, falling to 3.7 percent of equity. Impossible as it may seem, in 1986, while these companies suffered due to the dreadful market conditions, Flying J experienced a banner year (the company's fiscal year ending on January 31, 1987). Through the sale of assets and improvements in productivity, Flying J had a 25 percent return on equity, and net income reached 2.1 percent after taxes, both peaks in over a decade. Owners' equity went up more than one-third to $28.2 million and the resultant debt-to-equity ratio at 1.7 to 1 was lower than the 1984 ratio of 2.5 to 1! By turning what appeared to be certain financial disaster into a tremendous windfall, Flying J proved that it had the management skills to be an industry leader.

How important was the Husky acquisition to the company's future? The North Salt Lake refinery alone has allowed Flying J to grow far faster than would have been possible otherwise. The refinery has never had an annual loss under Flying J management, and it kept the company out of the red during the costly start-up of its massive retail interstate network. In fact, refinery profits have more than paid for the entire Husky acquisition plus all refinery capital improvements. Husky's 400-dealer network did not prove as valuable since dozens of operators dropped their affiliations or purchased their stations. They assumed Flying J was too small and too unknown to make it, and some resented losing their Husky name and ties.

The remaining dealers and franchised truckstops gave Flying J a considerable boost, although many continued to operate as standoffish independent owners. Of those truckstops that became directly owned by Flying J, most needed considerable refurbishing. Nevertheless, adding RMT extended Flying J's truckstop presence into 13 western states and tripled the company's outlets. With a highly profitable refinery and a nationwide network starting to take form, Jay and Phil were confident their retail game plan could be brought back on schedule. As Phil said, "Acquiring Husky was the greatest thing that ever happened to us." Buzz's observations are similar: "The Husky acquisition got us to where we are today."

EXPERIMENTING WITH FRANCHISING

*"Never become mired in defending an unworkable idea out of
some misguided ego or machismo motivation.
Cut your losses and move on."*
—James C. Freund

*B*eing deeply involved in the Husky takeover in the fall of 1985 did not deter Phil and Jay from starting to seek separate financing for travel plazas—a challenge that in its magnitude overshadowed any financial problem the company had faced to date. The numbers are simple to calculate. To have a network where truckers could travel on the interstates anywhere in the nation by fueling only at Flying J stations was estimated to require a minimum of 250 plazas. In 1986, with the average new plaza priced at $4 million, $1 billion in capital would be required. Flying J's shareholders' equity of $28 million in the same year was clearly insufficient to gain serious attention from lenders, especially with long-term debt already at $47 million. The company's financial position in seeking the required capital was like someone with maxed-out credit cards trying to buy a top-of-the-line Mercedes.

Phil and Jay's goal was to tie up the truckstop market as rapidly as possible. With the interstate system almost complete, a fixed number of favorable sites existed at the major interchanges, and these would be exceedingly expensive and hard to acquire in another decade. In addition, lurking in the background were the dozen or so giant petroleum companies flush with capital that could move ahead anytime they desired. (At the time, petroleum companies led all U.S. industries in sales and comprised one of the three leading industries based on total assets.) Fortunately, to this point, the hidebound petroleum titans stayed set in their ways and continued to focus on exploration, production, and refining where they enjoyed high profits and stellar reputations. Besides being disinterested, they were ill equipped to compete in the retail end of the travel plaza business.

As a reflection of their extreme optimism, Phil and Jay hoped to wipe out the competition by constructing 50 plazas a year for the next five years, thereby reaching the 250 goal. As it turns out, the assumptions underlying their timetable were incorrect but had few repercussions. They could not acquire the land, obtain the permissions, construct the plazas, nor put the financing in place to get the head start envisioned. However, this did little harm because it would take several years before competitors began to take serious notice. Jay and Phil were correct in that, at some point, the slumbering petroleum dinosaurs would awaken and become cognizant of the truckstop explosion taking place on the interstates, but none had the agility or knowledge to jump immediately into the fray.

When Phil took over retailing, he knew from his financial background that some novel strategy must be concocted to attract large backers. As he stated, "I liked to tell people we had the ideas, locations, people, and marketing to make it work. What we didn't have was the billion dollars to do it." He realized that such a proposal would be difficult to sell to Wall Street bankers and major brokerage firms because oil was considered a stagnant business, and the truckstop industry was still looked on as an ugly duckling. Going public with company stock to gain capital would also be difficult, and besides, Phil and Jay were adamantly opposed to giving up equity by selling stock to outsiders.

Seizing the Opportunity to Franchise

Anxious to get ahead with their plans on whatever basis possible (other than a stock issue), they looked into franchising. Franchising offered the quickest means for Flying J to get the large format plazas constructed under its flag. The franchiser did not have to put up the capital, thus relieving Flying J of its financing dilemma. Jay had looked into franchising as early as 1982. However, under the typical arrangement at that time, franchising would not have been very attractive because Flying J had only one small refinery, and franchisees would have to obtain fuel from other sources. Flying J's only benefit would be from the fees it charged franchisees, normally a fixed percentage that varied by type of product sold. Besides, the Flying J logo would only attract customers in a few western states.

In the early 1980s, franchising was commonplace in the automotive, transportation, and retail fuel industries. Hertz adopted such a plan in its rental car business beginning in 1918, and in the 1950s, several major oil companies (such as Conoco) constructed service stations and then leased them to dealers under a franchise arrangement. Of course, the majors had the common aim of selling dealers fuel, a commodity Flying J did not possess in large quantities. By the 1980s, the approach to franchising filling stations had changed little except for facility ownership. Under the most recent arrangement, a franchisee financed the construction and then operated the facility under a contract with a prominent brand-named fuel supplier.

In 1985, Jay was introduced to a franchising scheme where plaza financing could be obtained from third-party investors. Using limited partnerships, these investors would then lease the finished facilities to a franchisee who would then operate the facility, pay rent, and share income with the partners based on a formula that varied by percent based on the type of product sold.

Flying J was the franchiser but could also be a franchisee (an arrangement called a company-op). Flying J could build and lease back the completed plazas by using limited partnership capital. Adding to its chain in this manner was the next best solution the company had to outright ownership. In addition, the partnership provisions gave the franchisee the option of purchasing the facilities after leasing for ten years.

Jay happened onto this financing scheme while in Boise attending a seminar on raising business capital. There he met a person associated with Morton H. Fleischer, president of Franchise Finance Corporation of America in Phoenix, Arizona. FFCA in just seven years had raised more than $1.2 billion for fast-food chains through 14 limited partnership offerings. These franchisers were some of the country's largest fast-food companies such as Burger King, Pizza Hut, and Wendy's. Phil and Jay saw in FFCA the opportunity to duplicate this performance with Flying J travel plazas.

What Flying J wanted to avoid was the menagerie of facilities associated with current truckstop franchising. The franchise facilities of companies such as Unocal and British Petroleum exhibited inconsistencies similar to what customers find in many hotel and

Jay and Phil when undertaking franchising.

motel franchised chains. The franchiser is too often willing to take almost any lodging property owner under its wing, and customers never know what standards to expect at different locations. Many times an owner became a truckstop franchisee by doing little more than paying a fee and agreeing to use the major's fuel in return for the right to display the logo or brand name. Under the FFCA approach, this would be avoided because all franchised plazas would be new or remodeled units and had to meet Flying J design and construction criteria.

With the Husky debt still hanging around its neck, Flying J eagerly pursued its only immediate alternative if the company was to keep pace with its interstate plans. Fleischer was not hard to convince, stating, "There are 3,000 truckstops out there and nobody has networked them like the fast-food industry." In his view, "Flying J is bringing to the interstate system what fast food brought to the restaurant industry—quality, consistency of service, and cleanliness." Accordingly, FFCA set about to raise $1 billion through limited partnership offerings for the completion of 200 to 250 travel plazas in seven years.

Flying J's strategy was clear from the start. As stated in the company's fall 1986 *Newsletter*, "Flying J's primary goal through franchising will be to do for interstate travel what McDonald's had done for fast food dining by developing a nationwide marketing format designed to deliver products and service in the quickest, most economic, yet profitable way." The plaza design called for first-class facilities with consistent signage for easy customer recognition. Each plaza was to include separate diesel islands, ample parking, a 24-hour restaurant, convenience and trucker stores, tiled private showers, enclosed phone booths, game rooms, broker offices, and motels when appropriate. Service, low prices, and cleanliness were to be company signatures.

Unfortunately, three major differences set apart a fast-food outlet from a travel plaza. (All three later created problems for Flying J franchisees.) First, the investment in building and starting a fast-food outlet was less, then about $400,000 versus the $2 million to $5.7 million Flying J projected for travel plazas. (The estimate varied primarily due to the size of the facility, price of the property, and whether it included a motel.) Second, with on-site facilities that could include fuel islands, pay stations, a convenience store, restaurants, and a motel, a truckstop is far more complicated to manage. Finally, with motor fuel being essentially the same product at all locations and price comparisons simple to calculate from large exterior signage, customers tend to be extremely price conscious. In comparison, a fast-food outlet can differentiate a product such as a hamburger in numerous ways (single burger, double burger, cheeseburger, Whopper, bacon burger, fiesta burger, etc.). And the price differential is not a few pennies like fuel but can vary from 79 cents to $6.50 or more depending on the type of hamburger ordered.

Raising Capital through Limited Partnerships

The initial FFCA prospectus—released October 10, 1986—targeted raising $50 million (50,000 units at a $1,000 per unit), although the offering could go higher if oversubscribed. An investor was required to buy a minimum of five units of the "Participating Income Properties 1986, L.P." Initially the securities were sold through E. F. Hutton & Company Inc., a brokerage firm with more than 6,000 agents. Later, Hutton became part of Shearson Lehman, and this company assumed brokerage responsibilities.

The details of the franchising arrangements required Flying J to approve applicants, give advice on construction, provide start-up support, train key employees, and share its business systems and management expertise. The franchisee was to pay Flying J an enrollment fee and royalty payments on sales. The limited partners through the "participating income" provisions were to receive a return from the franchisee based on invested capital plus a percentage of sales depending on type (fuel, food, lodging, etc.). The minimum guaranteed return to the partners was 10 percent annually.

The initial offering raised approximately $50 million in 1987. Minus commissions, fees, and organization costs, approximately $44 million remained for "purchase of new and existing" Flying J travel plazas. Of the 11 travel plazas financed by the initial offering, independent parties leased only three. The others became Flying J company-ops. These were new units under design or construction, former Husky truckstops to be rebuilt, and already completed plazas (such as the one at Post Falls) that were, in essence, being refinanced.

When investors rapidly bought out the first partnerships, a second offering for $50 million was issued in 1988. This time, due to oversubscription, the total raised escalated to $82 million ($71.5 million net). This capital was used for 13 more plazas—9 company-ops and 4 new franchisees. Flying J, obviously eager to get its hands on financing, was the primary beneficiary of the first two offerings, gobbling up two-thirds of the capital provided.

A third offering of $50 million was released in early 1989. However, the glow was rapidly fading from this form of franchising when midway through the year, each of the seven independent franchisees was experiencing operating losses and, before year-end, all seven either sold or switched their leases to Flying J.

Given this bewildering experience, early in 1990 Flying J and FFCA agreed to discontinue the partnerships and drop franchising. Thus, the third offering was limited to approximately $22.5 million, making a grand total of more than $137 million raised to refinance, equip, build, and/or remodel 30 travel plazas or portions of plazas such as the motel added to the Boise truckstop. Including administrative and selling fees, the average cost per property was approximately $5 million.

Although its dream of obtaining $1 billion went unfulfilled, the net of $137 million came at a critical time in Flying J's history. Bolstered by this financing, Flying J at the close of 1989 had 38 plazas—24 company and 14 franchised facilities—enabling it to skip over several competitors and become number two nationwide in the truckstop/travel-plaza industry.[1]

---

1. FFCA leases provided the financing for most of the 24 company-operated plazas. The 14 franchised plazas were primarily former Husky stations owned or leased by parties other than Flying J. Some were independent FFCA-financed franchisees that had not yet turned over their properties to Flying J.

Key Provisions of the Partnership Arrangement

The legal and operating details dictating the relationship among Flying J, the franchisees, FFCA, and the limited partnership investors are too involved for detailed consideration here. However, to appreciate why the seven independent franchisees failed one needs an understanding of the key provisions defining the obligations of each party.

As noted, Flying J was to approve applicants, evaluate sites, and oversee design and construction. In relation to its business systems, Flying J required that the franchisee use its electronic point-of-sale system (recording and accounting for all sales transactions at a fuel desk). In addition, the company made available its inventory control forms and other business aids. The franchisee was to pay Flying J a $25,000 to $100,000 fee that in most instances guaranteed the recipient an exclusive territory, use of the Flying J name, and access to the franchiser's support services. Royalty payments to Flying J were set at 5 percent on gross nonfuel revenues and six-tenths of a penny to one cent per gallon on fuel. Monthly sales for a plaza were projected at between $800,000 and $1.5 million.

Capital from the limited partnerships was for equipping and building facilities that were then placed under a ten-year lease for each franchisee. After ten years, the franchisee had three options: back out, extend the lease for ten more years, or purchase the plaza at fair market value. The limited partners were to receive a percentage of all sales—3.5 percent for convenience stores and restaurants, 0.2 percent for fuel, and 8 percent for lodging. Being a "participating income" limited partnership, the partners (investors) were to receive annual rents and income from each property (Flying J's or a franchisee's) equal to a guaranteed 10 percent minimum on invested capital.

How the Franchisees Fared

Both Flying J and FFCA were breaking new ground with this form of partnership by applying it to truckstops. Accordingly, some problems were inevitable. (Limited partnerships have since gone out of favor and some former tax deferral features are now prohibited by law.) However, no one expected that within three to six months after the seven independent franchisee openings, each would experience significant losses and eventually fail. As so often happens with new businesses, the franchisees did not have the working capital or credit lines to sustain them until they could break even. Thus, they all ended up arranging with Flying J to take over their properties. The contractual agreement did not require Flying J to assume the lease if a franchisee wanted out, but Phil and Jay considered the FFCA franchised properties the equivalent of their other plazas and thus found no reason why they could not be just as profitable. Two of the most dissatisfied franchisees initiated lawsuits that were not settled until late 1995. The most common complaint by the independent franchisees was that Flying J had given them overly optimistic sales and profits forecasts. However, since most of the franchisees failed to follow Flying J's pricing and service guidelines, some shortfall in sales volume was inevitable.

When the franchisees' losses began to mount, some claimed that Flying J did not do sufficient hand-holding. The company responded by guaranteeing franchisees a six-cent difference between their selling price and what Flying J would charge for fuel if Flying J became their supplier. Thus, if the franchisees sold fuel at market that was only two cents above wholesale, Flying J would make up the four-cent difference, thus taking a hit on such deliveries. However, even with this backing, those franchisees that accepted this option continued taking losses and proceeded to appeal to Flying J to take over their properties.

The franchisees' problems provide textbook examples of the two most common causes of small business failure: underfinancing and weak management. Most independent operators were not sufficiently experienced or financed to withstand a major competitive struggle, one that inevitably occurs when a respected rival enters a new sales territory. In retailing, "heating up the ads" by cutting prices is a common practice used by competitors the day a new-

comer has a local grand opening. Obviously, neighboring companies are not going to sit by and lose customers without a fight,
knowing that single-business owners often lack "staying power,"
and that a prolonged price war could squeeze them out.

Several reasons account for the franchisees being underfinanced: some had apparently misrepresented their financial status
(Flying J could have screened applicants more thoroughly), and
working capital requirements were significantly underestimated.
In addition, in an industry noted for low margins, franchisees
found it difficult to pay royalties to a franchiser and rents to participating partners and still make a profit. However, as Flying J
demonstrated after taking over the properties, the main problem
was that the new franchisees, eager to get short-term results, were
unable or unwilling to accept Flying J's pricing philosophy. Thus,
they failed to generate projected volumes. With Flying J's experience, pricing philosophy, and deeper pockets, the company (after
assuming the leases) had each plaza generating a positive cash
flow, often within a few months. One plaza in the Midwest was
not only immediately profitable, but it became a company pace-
setter, being one of the first to reach monthly sales of a million
gallons.

In general, all original Flying J company-ops from the start
did well with less than 10 percent having losses extend into the
second year. One of the more successful was the Ehrenberg,
Arizona, plaza on the eastern border of Southern California. The
manager and half owner was Jay's son, Thad. For the next decade,
Ehrenberg became a model for others by being one of Flying J's
leading plazas in sales and profits.

A closely related problem hindering the franchisees was that
three of the seven independent franchisees (those in Dillon, South
Carolina; Knoxville, Tennessee; and Graham, North Carolina)
represented the first Flying J stations east of the Mississippi River.
Accordingly, these plazas gained little benefit from Flying J's name
recognition or its customer base except from transcontinental
truckers. In any retail industry, when a new company opens without a built-in customer base, it often takes a year or more to break

even. In addition, two plazas owned by independents had unusual size and elaborate design and décor, making their fixed costs higher than normal. Both exceeded the maximum franchise estimate of $5.7 million for construction and start-up costs, making it difficult to obtain a reasonable return on investment.

Another difficulty hindering franchisees was that some lacked the experience (and perhaps the desire) to run what is widely known as "a hands-on, 24-hour-a-day business," one that requires diligent round-the-clock attention by the station manager. The franchise owners were too often investors rather than on-site operators, which is recognized as the major cause of failure in all franchising regardless of the industry.

The importance of experience is clearly evident when comparing performance of the former Husky franchisees with the independent operators. All four Husky franchisees succeeded under the limited partnership arrangement whereas all independent franchisees failed. Of course, upgrading a former Husky facility required approximately half the investment of a plaza built from scratch.

In several respects, Flying J should have anticipated such problems. In the past, they had sold facilities to inexperienced owners and within a short time most failed. A prime example was the Flying J motel and station in Carson City, Nevada, where the new proprietor eventually declared bankruptcy. Most new owners (cautious and insecure because of their large investments) are eager to gain back part of their original outlay, and thus they raise prices seeking to boost revenues. The effect is the opposite: volume drops and net revenues generally decline. Flying J assumed that by providing franchisees its proven business systems, training them in plaza management, and urging them to adopt their pricing guidelines, they should be competitive. Unfortunately, most chose to operate differently, and Flying J had no legal basis to interfere. With the typical failure rate in franchising being less than 35 percent within four years, the demise of Flying J's independent franchisees must be largely attributed to the inexperience of all parties involved.

Favorable Outcomes for the Limited Partners

While franchising was disappointing for the franchisees, it was just the opposite for FFCA and the limited partners. In the first few months (less than a full year) when partnership capital was being deployed, the return to partners was only 9.5 percent. However, from the time all available funds were invested and Flying J took over operation of the plazas, annual partnership rents have never been lower than 11 percent and were generally nearer 14 percent. Important as the financing was to Flying J, the company paid a premium for this capital, especially in the era of lower long-term interest rates—ranging from seven to nine percent—that followed. Thus, soon after the ten-year leases were up and the facilities were open to purchase, Flying J began buying back some equipment and plazas. In March of 1999, the company (through a series of transactions) arranged with FFCA to gain ownership over all remaining properties by switching the remaining leases to mortgages payable over the next 15 years.

Some feel that franchising gave Flying J a "black eye." However, it was the only viable financing opportunity at the time, and most Flying J executives consider the outcome as positive, much like the questionable acquisition of Thunderbird. Though costly, franchising pushed the company ahead financially. Twenty-seven stations were added to the interstate network when otherwise the company could have financed at most seven to nine. Phil called the experience "painful and expensive . . . but we wouldn't be where we are today without it." He feels that Flying J learned two primary lessons: restrict partners to those who have your same goals, and accept partners only under circumstances where you keep control, especially in setting prices. Regarding his comment on goals, Flying J's were long-term: surpass the competition and develop an interstate travel plaza network that would put the company in the forefront of what was then considered an annual $26 billion industry. The new franchisees' concerns were more narrow and short-term: try quickly to show a profit by generously marking up prices and avoid any action that would threaten financial security.

Current Franchisees

As of this writing, Flying J has 15 stations franchised by affiliates who are independent operators. Eight are former Husky franchisees (primarily in the state of Washington) that have been with Flying J for 17 years. The ties between Flying J and the franchisees are simplified since 9 of the 15 are under control of Broadway Truck Servicing in Spokane, Washington. In 1994 Flying J constructed a plaza in Maryland that is now franchised to Gary Tsushima, a former Flying J employee who had 15 years of experience with the company including that of being a plaza manager. All 15 franchisees own rather than lease their properties. As franchisees, they pay a percent of sales to Flying J with no requirement to purchase fuel from the company. However, they must market fuel that meets Flying J specifications.

TAKING ON A
FINANCIAL PARTNER

"My greatest inspiration is a challenge
to attempt the impossible"
—Albert A. Michelson

*I*n 1990, the two major barriers preventing Flying J from moving forward with its plans for 250 interstate travel plazas still
remained: getting a guaranteed nationwide fuel supply and gaining the capital to construct 150 to 200 more units. The Salt Lake
refinery was limited to supplying company C-stores and plazas
within a 400-mile radius, an area including all of Utah and parts
of Wyoming, Idaho, Arizona, Nevada, and Colorado. With most
of the company's new facilities being built outside this perimeter,
obtaining contracts with reliable suppliers who would give prefer-
ence to Flying J during shortages or other emergencies again
became an issue. In these outlying regions, Flying J had relied primarily on the spot market (surplus fuel a refinery sells by bid after
meeting its contractual commitments). Spot-market sales are often
lower in price, but retailers cannot assume these leftovers will
always be available.

On the upside, at the end of 1990 Flying J's financial status was considerably improved compared with 1986 when franchising got underway. In these four years, all vital signs reflecting the company's financial health had nearly doubled: sales were up 75 percent (from $339 million to $592 million), net profit after taxes up 71 percent (from $7 million to $12 million), stockholders' equity up 77 percent (from $28.2 million to $50 million), and assets up 95 percent (from $107.4 million to $209.9 million). Profits as a percent of sales were a respectable 2.0 percent for both the beginning and ending years. As one would assume, because of this growth long-term debt also nearly doubled, going from $47 million to $88.9 million, a gain of 89 percent. Thus, the debt-to-equity ratio was slightly higher than earlier, remaining a formidable 1.8 to 1.

During this four-year span, Flying J eliminated five C-stores, dropping back to 51. Reflecting company strategy, the big gain—one that placed Flying J among the retail petroleum leaders—was in large travel plazas that jumped from 11 to 43 including franchised stations. The five new ones in 1990 (Reseda, Georgia; Walton, Kentucky; Troutdale, Oregon; San Antonio, Texas; and Rock Springs, Wyoming) were all financed through limited partnerships. Surprisingly, as partnership financing ended, Flying J's design and construction efforts expanded rather than declined with 16 new sites under development at the end of 1990. Of these, 9 were completed and opened for business during Flying J's next fiscal year ending January 31, 1992. Reflecting the nationwide emphasis, one each was built in Arkansas, California, Indiana, Montana, Nebraska, Nevada, Ohio, Texas, and Virginia. With this expansion, in 1989 Flying J finally saw retail sales exceed those from upstream operations (refinery, oil and gas production) and wholesaling.

This upsurge in construction kept the company under pressure to gain needed financing as the limited-partner dollars were quickly running out. Of the 16 plazas under development at the 1990 year-end, the only 3 being financed with limited-partnership capital were those in Bakersfield, California; Winnemucca, Nevada; and Wythville, Virginia. With still less than one quarter of the projected 250 plazas in place and another 16 in design or construction, Flying J was forced to intensify its eight-year hunt to tie down the necessary financing.

Considering a Financial Partner

In 1989 when it became apparent that FFCA franchising was not the answer, Jay became more intrigued with the idea of adding a partner who could supply capital and, in the process, reduce Flying J's projected debt. With a continued reluctance to give up part of his company by selling stock or merging, Jay liked the idea of a joint venture. Both parties would share in plaza ownership but each would retain its own corporate identity. If the right petroleum partner could be found, Flying J could solve both its supply and financing problems. Such a joint venture would be especially ideal if Flying J could continue to capitalize on its main strengths: plaza design and operation. An added benefit from joining with an established company is that, at last, Flying J would obtain the credibility needed to overcome the concerns of investment bankers, commercial bankers, and insurance companies that held most of Flying J's long-term debt.

Appealing as the idea was, Jay kept it to himself because of the fear that employees might think he was trying to sell the company. Before long, others in Flying J were also coming to conclude that an outside partner was the only solution. Phil, ever anxious to move up and become an industry leader, viewed with envy the strong balance sheets of the major petroleum firms. Buzz, as expected, was interested in getting a secure source of fuel from a nationwide supplier. The petroleum magnates were also starting to take notice of this upstart company in Utah. At this time, Flying J was far from being the top national retailer in diesel sales, but its rapid growth in monthly gallons sold had to catch the eye of major refiners.

Searching for a Financial Partner

Finding a financial partner took several steps. As noted, Flying J in
most of its history had relied on the spot market to obtain fuel,
primarily because of the lower prices. However, the prospects of
guaranteed contracts to reduce risks had intrigued the company
since the shortages of the mid-1970s. In addition, as Flying J's
sales continued their remarkable upward climb, it put the company in a better position to secure price advantages by concentrating purchases among two or three majors, one being Conoco Inc.
At the time, Conoco refineries were often operating at less than
capacity and had excess fuel in storage. Conoco also would benefit from avoiding the erratic spot market, and thus the company
showed an interest in joining with Flying J in a long-term fuel
agreement, a natural marriage for both parties.

Using a subtle strategy he had often employed, Jay did not
directly approach Conoco or any other major with an equity proposal. His first step was to contact the major brand name companies in search of more long-term supply arrangements. Conoco
was the first company to show interest due to its current favorable
relationship with Flying J and to the flat current international
demand for oil. With Conoco paying more attention, Jay and Phil
decided to expand their search and see if other majors could be
enticed into a more attractive supply or possibly equity deals.
Accordingly, the company sent inquiries to ten or so big name
brands. The letter stated that Flying J was building a national network of plazas and hoped to be the largest diesel marketer in the
country. Accordingly, the company was seeking an agreement with
a major supplier who would make a long-term commitment as a
financial partner. Reacting in their traditionally conservative manner, most majors contacted showed little interest or were slow to
respond. They viewed this brash upstart as doing a little "puffing"
and unlikely to deliver on its forecasts.

At the time, conversations were continuing with Conoco officials. Discussions moved from a solitary supply agreement to their
taking an equity position in all travel plazas. Conoco was soon

attracted by more than just harnessing its largest diesel customer. In the 1980s Conoco had purchased a dozen truckstops from Marathon Oil and was now operating 17 including two that it had recently constructed. Much to the company's despair, it was losing rather than gaining market share. As one Conoco representative stated, "Our stations paled in comparison to those of Flying J." Thus, Conoco executives decided that a closer alliance with a company displaying the potential to become the preeminent national truckstop operator would be in their favor.

When it became evident that Conoco might come on board under an equity arrangement, Jay invited their representatives to Brigham City to view Flying J operations and discuss possible alliances. As Phil stated, "We spent a couple of days with Conoco representatives, had dinner at Jay's house, followed by a couple of more days on the road looking at facilities." This led to consideration of a partnership arrangement. Spurred on by their extremely favorable ties in the past, the talks between Flying J and Conoco officials culminated in a joint venture. The announcement of the new company, known as CFJ Properties, was released on January 9, 1991. The joint venture was formally established under Utah law as of February 1, 1991.

In the press release announcing the partnership, each party acknowledged certain benefits. Phil was pleased to note, "Conoco brings to the partnership, not only its well-founded experience in the oil industry, but a brand name recognized for quality products." Rick Hamm, general manager of Conoco's branded and retail marketing operations, added, "Flying J is a major Conoco customer. They have developed innovative marketing programs for travel plazas, and their reputation for personal service has made Flying J the leader in the truckstop industry." Both executives agreed that the joint venture would be long term, and together they would be aggressive in expanding Flying J's current interstate travel plaza network.

Those within Flying J and many outsiders considered the partnership as the last piece of the puzzle needed for the company to gain mastery over nationwide retail diesel sales. They viewed it

Flying J and Conoco representatives at the time of the agreement.
Left to right: Barre Burgon (FJ), Phil Fredricksen (Conoco), Ed Adwon
(Conoco), Jay Call (FJ), Rich Hamm (Conoco), Phil Adams (FJ),
John Barr (Conoco), and Buzz Germer (FJ).

as a sacrifice in ownership, but as one observer stated, Flying J would now be able "to get to its goal of being the interstate travel plaza leader twice as fast." Jay called it a "wonderful deal for us and for me personally. It took off the financial strain, and it enabled us to move fast enough to capture the marketplace the way we needed to."

Outsiders were again flabbergasted at what Flying J had accomplished. One corporate leader called the formation of the partnership "absolute genius." Another major Utah petroleum wholesaler when referring to Jay stated, "Anyone who could pull off a coup like he did with a major oil company—that is unheard of. Major oil companies are very secretive and very careful about getting involved in those kinds of ventures."

The agreement immediately took financial pressure off Flying J's pocketbook. As one bank president stated, "Our bank examiners often wanted to downgrade Flying J loans, but the Conoco agreement eliminated most of that." In general, Jay, Buzz, and others agreed with Phil who concluded that Flying J might have moved ahead on its own, but it would have been more expensive and drawn out, and caused "a lot more sleepless nights."

What Conoco Brought to the Table

Like Flying J, Conoco had its beginning in Utah. Isaac Blake, who earlier had been involved in marketing petroleum in Pennsylvania, came to Ogden, Utah, in 1875 with the intent of using this city's railroad hub to market petroleum products (mainly lubricants, harness and carriage oil, and paraffin for candle making) through-out the West. Along with other investors, he established Continental Oil and Transportation Company. Two years later, Blake was instrumental in founding the Continental Oil and Transportation Company of California, a company that gained recognition for transporting oil products in tanks on horse-drawn wagons and on railroad cars. In addition, this company gained prominence in California through establishing pipelines to transport liquid petroleum products.

After the company in Ogden launched numerous branches in the Rocky Mountain region, Blake established the Continental Oil Company of Colorado with headquarters in Denver, leading to Continental becoming the dominant petroleum company in the West. The rise of Standard Oil in the eastern states led to an inevitable struggle for supremacy between the two combatants. Standard Oil, bolstered by the Rockefellers' questionable practices, gained the upper hand, and Continental Oil Company became an affiliate of its much larger competitor on January 2, 1885. Eventually Standard Oil under John D. Rockefeller came to control 90 percent of the refining in the United States, causing the government to apply antitrust laws against the company. In 1913 the Supreme Court required Standard Oil to divest many of its holdings, and Conoco became independent again. In 1929 Conoco merged with Marland Oil, a large oil producer headquartered in Oklahoma. In the 1950s Conoco gained international stature by making major oil strikes in the Middle East, enabling the company to take its place among the principal domestic integrated petroleum corporations.

In the years just before the formation of CFJ, Conoco maintained annual national rankings of between 12th and 15th on each measure used to compare the size of the integrated petroleum giants (crude oil production, refinery capacity, and gasoline motor sales). Conoco's refineries supply customers in most of the U.S. except on the West Coast and the extreme Northeast. Conoco, an industrial power on its own, gained formidable financial security in 1981 when it became a subsidiary of DuPont—a company ranked as the 13th largest domestic corporation by *Fortune* magazine. With a rock-solid financial foundation, Conoco was an extremely attractive partner for Flying J.

Terms of the Partnership Agreement

Thirty-three facilities immediately came under the arrangement. The 50-50 partnership required that each party initially put in $45 million, mainly in properties. As part of the deal, Conoco's share involved the purchase of ten Flying J plazas. Flying J contributed the other 23, all leased plazas financed by FFCA. Through the agreement, Flying J gained not only a large influx of capital, but also control over plaza operations and the right to handle the associated trucking. Recognizing Flying J's skill as an operator, Conoco did not object. All direct expenses were to be accounted for in the partnership, and Flying J was to be reimbursed monthly for CFJ's share of projected corporate overhead. The agreement specified that both names be on highway signage with Flying J's listed over the diesel islands and Conoco's over the gasoline lanes.

A six-person executive committee manages the joint venture, three from each partner. The committee is concerned primarily with approving plaza design, financing, evaluating market potential, and other similar long-range concerns. Short-range topics center on the performance of each plaza, litigation, bonuses, and profit-sharing contributions. Besides the 33 plazas then in place, eight more under design were part of the initial agreement. Both parties have the right to be a full partner in all future plazas. Accordingly, joint ownership may include on-site C-stores, service centers, restaurants, and lodging. This right relates to plazas under consideration by either company. Until 1998, Conoco rarely declined participation with two exceptions: when a proposed plaza would likely damage business at an existing Conoco retail outlet or when smaller "fuel stop" facilities were under consideration. The partnership is restricted solely to interstate operations. Flying J's 59 C-stores, four other motels, the refinery, and production properties were excluded. In addition, all franchising is handled by Flying J outside the partnership.

Outcomes of the Partnership Arrangement

In general, both parties have been satisfied with the arrangement. Flying J found a stable, financially strong partner who would keep to the terms of the original agreement. Flying J leaders now felt they could fulfill their dreams (many considered a fantasy) of building a travel plaza per month until a motorist could circle the United State and never fuel other than at Flying J facilities. To Jay, this alone made the entire partnership worthwhile. But the partnership gave Flying J another significant financial break. Under the FFCA contractual arrangement, Flying J was paying the limited partners close to 14 percent for its leases. With Conoco as a partner, travel plaza capital could be obtained at 7 to 9 percent.

Conoco realized similar major benefits. Three years after the partnership was formed, company representatives referred to it as the "shining star" within Conoco. Ron A. Sumner, Conoco's director of joint ventures, claimed to be "the luckiest person in Conoco to be associated with this venture." He added, "The partnership is profitable, growing, and dynamic, and it is changing the industry." At the time, Conoco's only complaint was that the joint venture was not building plazas fast enough. Unfortunately, the permitting process (getting various zoning approvals, water rights, sewer connections, etc.) often took two or three years, and some cases were still pending after five years. Conoco is generally the more conservative partner, taking a go-slow, wait-and-see attitude on ancillary facilities sometimes proposed such as a motel or a J-Care service center for trucks. With the former, Conoco is especially cautious since it has had no experience in managing lodging.

Conoco has been an ideal partner by staying out of the details of plaza operations, deferring to Flying J as the master operator. Initially Flying J had some uneasiness about Conoco's intent in forming a joint venture. As Barre Burgon stated, "We were a little concerned up front that the big oil company wanted to get its nose in and we would be out the door." The result has been just the opposite. As Sumner stated, "They had the operational experience, and we agreed we would not get into the day-to-day business, and, for the most part, we've stayed out of it." Being the larger partner, Conoco has not tried to muscle its smaller counterpart.

Perhaps Flying J's primary difficulty in dealing with Conoco is reconciling a small, agile company's ability and willingness to make quick decisions with a more bureaucratic counterpart used to conducting numerous studies and waiting to gain approval of a mixture of committees and higher-level executives before going ahead. As one Flying J employee said, "They take decisions up their ladder, and it can take a long time to get a decision. It's hard for Flying J to live with that. We have always been a lean machine."

In addition, the two parties experience some dissonance due to their different goals. Following its long-held practice, Flying J's primary interest is in building future value rather than in obtaining more favorable short-term results. Thus, the company remains as leveraged as possible, ignoring the financial dangers involved. On the other hand, Conoco, being a public company, is eager to give shareholders more immediate returns and minimize those risks that appear to be extreme. As a result, recently Conoco has declined participation more often than in the past. Of the 80 Flying J plazas built between 1990 and 1999, the CFJ partnership owns 55. Overall, it has been rare to find such compatible business bedfellows.

What the future holds for the partnership is difficult to predict. The CFJ joint venture has come to dominate Flying J's primary business. In 1999 the partnership accounted for 90 percent of fuel sales. The number of plazas has grown to where in some regions Conoco cannot supply as much of CFJ's fuel requirements as they would like. Initially Conoco provided approximately two-thirds of the partnership's fuel, representing 40 percent of Conoco's total diesel sales to all customers. However, with more than 50 CFJ plazas added between 1994 and 2000, Conoco's distribution network and refining capacity can now meet only 50 percent of the partnership's current fuel needs.

Conoco's Ties with DuPont

In 1981 Conoco was acquired by Du Pont (E.I. du Pont de Nemours), the largest domestic chemical company. This was just after the oil shortage scare of 1980, and DuPont bought Conoco to assure itself of a steady oil supply for its large line of petrochemicals including nylon, Dacron, Teflon, and Nomex. At the time, DuPont, a Goliath in the domestic chemical industry, had twice the annual revenues of any other domestic chemical company and, as noted earlier, was consistently ranked among the highest 15 U.S. corporations based on revenues.

When oil prices hit a ten-year low in 1998, DuPont found similar reasons to spin Conoco off. Although its petroleum subsidiary had provided 50 percent of DuPont's profits in 1997, low oil prices made petroleum companies unattractive. Moreover, DuPont, under a new CEO appointed in February of 1998, was starting a wholesale redesign of its product lines and image. The company began to restructure by emphasizing research and products in the life sciences (agriculture, pharmaceuticals, and health and nutrition products). Needing more capital to acquire life sciences companies and to reinvent itself, DuPont tried to peddle Conoco at an attractive price. After failing to do so, in May of that year the parent company decided to divest 30 percent of its Conoco shares through an initial public offering (IPO). On

October 22, 1998, DuPont offered \$4.2 billion in Conoco shares at \$23 a share. The price on the exchanges rose to \$24 7/8 before the trading day was over. At the time, it was the largest domestic IPO ever and a huge success for DuPont.[1] Following the masterstroke of the initial IPO, DuPont spun off the remaining 70 percent of its oil subsidiary by giving DuPont shareholders the option to swap its stock for Conoco's. The tendered shares far exceeded the supply, resulting in less than half of the shares tendered being converted.

Being Independent and Anticipating a Merger

In early 1999, almost immediately after becoming independent, Conoco (following the lead of most major petroleum companies) announced that it was cutting staffing and reducing expenses in the wake of crude prices dropping to an unprofitable \$12.50 a barrel in 1998 and the first quarter of 1999. Conoco intended to eliminate 975 jobs as part of its aim to slash expenses by \$500 million. The announced reason for the change was "to position the company to deal with the lingering problem of low crude prices."

Initially, being independent had not harmed Conoco. While part of DuPont, it maintained an acceptable profit record for its industry, although earnings growth was essentially flat from 1994 through 1998. Though disappointing, it still outperformed other companies in the same industry that together experienced an average earnings decline of 2.1 percent. During 1999, when crude prices more than doubled after the first quarter, Conoco's returns were far better than its former parent's. For the year, Conoco's earnings went up 65 percent whereas DuPont's dropped 87 percent. In 1999, Conoco's sales were \$27.3 billion, nearly ten times that of Flying J's. In 1999, based on annual sales, Conoco was ranked by *Business Week* magazine as fourth in the domestic fuel industry (coal, oil, and gas).

---

1. The United Postal Service (UPS) IPO in November 1999 was the first to surpass the Conoco IPO, raising \$5.4 billion. AT&T Wireless and Kraft foods followed.

Probably the primary benefit Conoco gained from its new-found freedom was being able to take advantage of investment opportunities. DuPont had annually siphoned off a significant share of Conoco's cash and had been restrictive in approving the company's requests for major capital improvements. This does not necessarily mean that Conoco would pour more capital into CFJ. Flying J remained a small portion of Conoco's total sales, and as an independent, Conoco was more eager to satisfy its stockholders and debtors by having commendable earnings, significantly higher than those obtained through CFJ.

Being more aggressive as an independent, Conoco on May 29, 2001, agreed to buy Gulf Canada Resources for $4.3 billion in cash, a move that would dramatically increase Conoco's international oil reserves. The addition of Gulf Canada's proven oil reserves of more than 1 billion barrels worldwide would jack up Conoco's reserves by almost 40 percent. The agreement became effective two months later when more than 95 percent of Gulf Canada's shares were tendered.

Within three months, Conoco was again on the move, announcing on November 19, 2001, a proposed merger with Phillips Petroleum Company. The $35 billion merger would create the nation's third largest integrated oil company with more than $60 billion in assets. The agreement to form the new company—ConocoPhillips—was jointly announced as a "merger of equals" although Phillips had 38,500 employees and $35.4 billion in assets versus Conoco's 20,000 employees and $27.7 billion in assets. If the merger is approved, Phillips shareholders would own 56.6 percent and Conoco shareholders 43.4 percent of the new corporation.

In explaining the reason for the merger, Conoco chairman Archie W. Dunham stated, "We think this is the best way to create long term and short term value for our shareholders." The move to combine the two companies was typical of the merger frenzy going on in the industry. The 1990 decade of poor performance for petroleum companies caused even the major producers, such as Exxon and Mobil, to merge with the intent of gaining efficiency, solidifying their financial status, and benefiting from economies of scale. Phillips and Conoco representatives stated that by merging the companies they should save at least $750 million annually. As one investment analyst viewed the announced Conoco-Phillips merger, "This is absolutely a matter of survival. . . . If oil and gas prices collapse, smaller companies will be swept away."

Worldwide, the joint corporation would become the sixth largest energy company based on hydrocarbon reserves and the fifth largest global refiner. Phillips has an especially strong retail presence. It sells fuel at more than 12,000 stations under brands such as Phillips 66, Circle K, and 76. Conoco operates 7,000 stations throughout the world.

The deal is expected to close in the second half of 2002 after obtaining regulatory approval. Shareholders of Phillips and Conoco overwhelmingly approved the merger in a stockholder meeting held March 12, 2002.

BECOMING THE PACESETTER IN INTERSTATE TRAVEL PLAZAS

*"The entrepreneur always searches for change,
responds to it, and exploits it as an opportunity."*
—Peter F. Drucker,
management statesman

*T*he 1990s proved to be one of the golden eras in U.S. economic history. The economy, bolstered by low inflation, experienced the longest period of sustained growth in modern times topped by an average annual increase in gross domestic product of 4.5 percent from 1995 through 2000. Such growth was double the annual average from 1973 to 1990. During the decade, national productivity expanded at an average annual rate of 2.4 percent. Productivity gains were spurred by investment in technology that increased at an inflation-adjusted 20 percent annual pace, becoming an unprecedented 35 percent of total capital spending. These conditions pushed the stock market beyond what even the most optimistic forecasters predicted, soaring 270 percent based on the Wilshire Total Market Index. Such gains created enormous wealth for investors and corporations, part of which vanished in the downturn beginning in the latter half of 2000 followed by the tragic events of September 11, 2001.

Most of the decade's productivity and economic output increases can be attributed to the growling engine of change—technology. In this instance, the engine was fueled by advances in electronic data processing and improved means of capturing the business potential of the Internet. Companies such as Microsoft, Oracle, Cisco, AOL, Verizon, and Dell Computer began appearing near the top of the Fortune 500 leader board, and some recently established high-tech firms temporarily passed General Motors in market value.

Beginning in the late 1950s, the computational power of the world went up 84 percent each year, growth never experienced in any other technology. However, it was not until the 1990s that corporations gained the full benefit of this runaway technology by incorporating it into virtually every aspect of their business operations. As an example of how the Internet boosts efficiency consider banking. The cost of a typical bank transaction by a teller is $1.25. This decreases to 54 cents by phone, and 24 cents at an ATM. The same transaction cost is a mere 2 cents on the Internet, and this does not take into consideration the lower associated capital and real estate investments. As another example, online stock trades cost 20 percent of conventional methods.

Although the big oil firms invested heavily in technology, to their dismay they did not share in the decade's economic boom until 1999–2000 when crude prices tripled. Hindered by global oil consumption plodding along at a 2 percent annual growth rate and falling energy prices cutting into corporate revenues, the petroleum majors suffered their worst decade since the 1930s. Major refining companies were burdened with a meager 4 percent return on capital that left the industry in disarray. At the beginning of 1999, just prior to crude prices tripling, one analyst wrote, "Oil companies are not earning their cost of capital. It's an industry in liquidation, basically." Reflecting the fickle nature of the petroleum market, after modest gains in the 1990s, in 2000 refining companies on average experienced better than a 50 percent improvement in revenues and a 148 percent gain in profits. That year, for the first time since 1984, Exxon Mobil replaced General Motors at the top of the Fortune 500 as the largest domestic corporation.

Contrary to the boom petroleum companies enjoyed, 2000 was devastating for those with transportation as a major cost component. While refiner profits were going up 148 percent, trucking company profits slid back to an average gain of 2 percent, resulting in many carriers suffering severe losses. In November and December of 2000, one report identified 2,300 trucking companies that filed for bankruptcy. The turndown took an estimated 10 percent of nationwide trucking capacity off the road. With fuel prices going up and demand going down, those truckers and carriers holding minimal cash reserves were hit the hardest. In the last six months of 2000, 50 percent of the 51,000 class 8 (18-wheel) trucks pulled out of service were those owned by independent operators.

From 1990 through 1999, petroleum retailers suffered the same stagnation as refiners and producers. Gasoline and distillate consumption increased at an average annual rate of less than 2 percent. However, the demand for on-highway diesel fuel used in truck transportation gradually increased during the decade, topping out at 6 percent growth in 1999 before dropping to 3.2 percent in 2000. When crude prices tripled, refined fuel prices did not increase proportionately, largely because crude oil accounts for approximately one-third of the price of gasoline. Thus, average domestic gasoline prices rose 28 percent—from $1.22 per gallon in 1999 to $1.56 per gallon in 2000.

Flying J in the 1990s

Given the decade's dreary conditions for petroleum companies (including retailing), how did Flying J fare? With the financing in place, an aggressive young president at the helm, and competitors slow to respond, rather than stagnating with the balance of the industry Flying J continued its remarkable climb to become the nation's leader in interstate travel plazas. In the process, it rose to 46th on the *Forbes* list of the largest U.S. private corporations, and 493rd on the *Hoover's* ranking of all private, public, and governmental business enterprises. During these ten years, the company tripled its plaza locations, growing from 43 in December of 1990 to 141 as of January 31, 2001. Of this increase, 79 were company designed and constructed, 4 were purchased from others, 4 were remodeled Husky franchises brought back into service, a net of 5 new franchises were added, and 6 "associates" came on board January 1, 2001.

The "associate" designation resulted from an agreement between Flying J and Golden Gate Petroleum of Oakland, California. Under the agreement, six of the local Golden Gate Cardlock stations switched to Flying J's electronic and communication point-of-sale systems. These lightning fast systems produce an array of valuable management information such as detailed reporting on credit, sales, and fuel entries. These data offer the ultimate in credit card management and in comprehensive feedback to drivers and owners. As a bonus, Golden Gate cards are accepted at all Flying J locations, and cardholders have access to Flying J's loyalty programs.

Flying J results in 2001 made it clear that the company strategy formulated in the mid-1980s was paying gigantic dividends. In the early 1980s, the last links in the country's interstate freeway system were completed, allowing truckers to concentrate their travel on these super highways. The interstate system, only 1.1 percent of national highway miles, now handled more than 60 percent of truck-trailer traffic. This opportunity to capture more of the trucking trade was also enhanced by the minimal services provided to professional drivers at that time. Estimates of the number of truckstops vary depending on the source used and the size of facility in the count, but somewhere between 1,100 to 3,000 were said to be along the interstates, with few (considerably less than 10 percent) being state of the art. Of these, some were one-stop centers where a traveler could obtain fuel, convenience store items, meals, and lodging, but none could match Flying J's quality. Truckers, most of whom spend three out of four weeks on the road, strongly appreciated Flying J's effort to make its plazas a "home away from home" for the often weary professional.

Flying J's vision of the national travel plaza market and the failure of competitors to respond are evident in the company's progress during the 1990s. While Flying J was completing 79 plazas, its three major competitors together opened a mere 65. The big oil petroleum companies continued to find retailing unattractive and those smaller companies currently in the truck-stop business lacked either the financing or leadership to move rapidly ahead. As Flying J's taillights grew dimmer to its competitors, near the end of the decade some reacted by attempting to gain market share through merging, gaining greater outside financial support, and/or speeding up the modernizing of current facilities.

During the decade the 83 plazas Flying J added (79 built plus four purchased) helped fill in some of the remaining links on its coast-to-coast network. With few exceptions, long-haul drivers could now travel throughout the nation and only rely on a Flying J credit card to cover all job-related expenses. Reflecting this continent-wide expansion, 33 of the new facilities were in the south/southwest states, 22 in the Midwest, 17 in the Intermountain area, five in the Northeast, four on the West Coast, and two in Canada. Current plans call for a network of 140 super plazas plus many more in the medium classification within four years. This scheduling could vary depending on the time required to obtain appropriate government permits, an especially time-consuming process in the Northeast and California. As one measure of how these new plazas enhanced the lives of truckers, during the decade Flying J added 975 showers and 12,000 parking spaces at stops along the interstates.

In the region surrounding Flying J's recently relocated home office in Ogden, Utah, few local residents are aware of the nationwide presence the company commands as shown in the map on the inside of the front cover. Local traffic flow is insufficient to justify large facilities in this section of the country where Flying J

C-stores had their start. Of the 19 facilities built in the Intermountain area during the last decade, the majority are medium-size or smaller units that rely on the North Salt Lake refinery for fuel. As a result, in 1999, Flying J ranked just tenth in Utah based on state taxes collected on gasoline sales for automobiles. With few imposing Flying J super plazas within a 300-mile radius of company headquarters, Utah residents are normally surprised when traveling out of state to see the quantity, size, and grandeur of the latest full-service plazas.

During the last decade, the company decided to essentially phase out of C-stores by selling or closing most of the 51 operating in 1980, though some were expanded into larger fuel stops by adding diesel islands. In October of 1998, the eight remaining C-stores (selling gasoline only) were combined with the refinery into Big West Oil LLC, a wholly owned subsidiary, with Buzz Germer as president. A key reason for reduced emphasis on C-stores is that this segment of the industry was rapidly becoming overcrowded, especially with hyper stores like Wal-Mart and Albertson's adding islands to sell gasoline.

Flying J's expansion during the decade has been especially rewarding to the professional driver. With the company being licensed to do business in 43 states and Canada, drivers gain the benefit of having the Flying J logo greet them on essentially every segment of the nation's interstate system. At each location, they can count on the same low prices, high standards of cleanliness, multiple appetizing food offerings, quick service, outstanding hospitality, a wide range of merchandise and gifts, and other amenities. Such consistent standards set Flying J apart from its major competitors. With more than 20 years of experience in the truckstop–travel plaza business, Flying J architects are continually fine-tuning plaza designs and updating aging facilities. In the goal to be the hospitality leader on the nation's highways, the company takes special care in selecting, training, and evaluating its employees.

Results from a Decade of Rapid Growth

> Driver comment:
> "I find the Flying J Travel Plazas to be the finest of *all* chain plazas. If I have a layover, I make sure that I find a Flying J. Here I can be assured of ample parking, a clean shower, and good food at the end of the day. Thanks for being there, Flying J!"
> —John Roden, Ludlow, KY

In the 1990s, the company's crowning accomplishment was to become arguably the nation's largest diesel provider to the trucking industry. Determining such positioning is often subjective and always difficult to calculate. With the variety of statistics available, someone willing to selectively "pick and choose" can back up almost any claim. Federal and state agencies collect data on truck and car registrations, demand and supply for petroleum products, highway statistics, fuel taxes, and a variety of other transportation components. A private organization, the American Petroleum Institute, does the same for the petroleum industry although its primary focus is on crude oil and refined products. The American Trucking Association gathers data of interest to its members, and the National Association of Travel Plazas and Truckstops (NATSO), a national trade association, collects numerous statistics on fuel retailing for its membership, which does not include all chains or owners (or Flying J). However, NATSO's industry statistics are the most useful because these cover the market that most nearly matches Flying J's product lines.

An example of the verbal confusion is the various ways "diesel" can be interpreted. Federal government data on "distillate fuel oil" include all that supplied to domestic residential and commercial users. Of this total, just over 50 percent is "low-sulfur diesel fuel" for trucks and cars, the measurement used to compare fuel sales in the travel plaza industry. The balance in federal reporting is diesel for trains, ships, farmers, construction, and the heating of homes and other structures. Figures on consumption also tend to be misleading because retail reporting often combines diesel and gasoline sales.

Given this confusion, Flying J's dominance in fuel sales can still be inferred from data available. The company sold approximately 2.5 billion gallons in 2000. The total for the country was slightly in excess of 33 trillion gallons. Thus, no company captured a major share of the market. Flying J's average super plaza monthly sales of approximately 2 million gallons are double the industry average for large "full service truckstops" as reported by NATSO, and Flying J has the largest number of these premium plazas. In addition, on the interstates the company outdoes competitors in gasoline sales. Gasoline comprises approximately 15 percent of Flying J fuel sales versus 10 percent on average for companies with comparable travel plazas and truckstops.

Financially, Flying J likewise experienced success during the last decade of the twentieth century. Its position in the industry is easier to document using financial data. The company continued on track with its remarkable percentage gains in annual total revenues and growth of assets at a time when its size meant that obtaining such multiples involved taking huge chunks of sales from competitors. During these ten years, sales rose nearly 460 percent, increasing from $776 million in 1991 (Flying J's fiscal year ending January 31, 1992) to over $4.3 billion in 2000 (year ending on January 31, 2001).[1] According to NATSO, industry sales for 2000 were $42 billion, giving Flying J more than 10 percent of the industry total, a significant share given the 600 or so companies operating plazas. The most amazing aspect of Flying J's financial showing is that it occurred when most petroleum company sales were inching along at no better than inflation.

To many it was unthinkable that Flying J could keep pace with its historic 20 percent average annual revenue increase, especially in calendar years 1997 and 1998 when growth dropped to single digits resulting from a dispute with Comdata, a third-party billing company specializing in the trucking industry. Comdata is the dominant credit provider and information processor serving

---

1. The increase is a more correct 440 percent when accounting changes are included such as the additon of sales taxes in revenues.

transportation in the U.S., and also owns the Trendar point-of-sale system that handles the bulk of industry credit transactions. By necessity, transportation has its own set of credit cards and processing methods. The services and reporting available through transportation industry cards are much broader than standard commercial cards such as Visa or MasterCard. In the instance of Comdata, its Comchek card and Trendar point-of-sale device, unlike those of Visa or MasterCard, allow a carrier to set limits on what is purchased and when and where the card can be used by drivers. The company can then arrange for periodic reports detailing these purchases by vehicle, driver, and other variables. Such reports are useful in cost control, financial analysis, payment of taxes, and so forth.

In 1995 Flying J started marketing its point-of-sale system (ROSS) to other truckstops followed by an announced intent to become a third-party billing company. This became a reality when Flying J issued its Transportation Clearing House (TCH) card that holders could use industry-wide. This card was made available to any carrier, hence putting Flying J in direct competition with Comdata and some other smaller third-party billing companies. When one of the larger truckstop chains prepared to sign on with Flying J, Comdata indicated that it would not process one of its direct bill products on the ROSS system, making it difficult for that company to switch. Comdata took other steps to discourage use of the ROSS system causing the prospective buyer to back out of its agreement. In addition, Comdata reacted to Flying J's introduction of the TCH card by threatening to raise the transaction fees it charged truckstops if they accepted this card. Because Comdata transactions constituted more than 50 percent of all credit transactions at most truckstops, these threats had the effect of discouraging proprietors from honoring the TCH card.

Beginning June 1, 1996, Comdata went one step farther by stopping the acceptance of its Comchek card at Flying J locations, and at non–Flying J locations, Comdata refused to modify its widely used Trendar machine to accept TCH transactions. Most companies, especially large trucking firms, were locked into

Comdata's reporting and credit operations. Accordingly, they had little choice but to instruct their operators to discontinue charging at Flying J facilities.

Flying J's obvious recourse in July 1996 was to bring suit against Comdata claiming that the company was violating antitrust laws by denying it equal access to the market—in effect, not treating it the same as other third-party billing companies. In addition, Flying J claimed that Comdata had unlawfully attempted to exclude Flying J from entry into the industry point-of-sale market. In May of 2001, this dispute was settled outside of court. According to Comdata press releases, Flying J received a $49 million cash settlement, and Comdata agreed (among other things) to modify its Trendar software to accept the TCH card. (As of October 15, 2001, Trendar began processing the TCH card.) The dispute temporarily stalled Flying J's growth curve and the company is still struggling to get universal acceptance for its card, but, as frequently occurs in battles for survival, those who endure eventually come back stronger.

Those assuming Flying J would be brought to its knees when it lost thousands of customers were shocked to see how the company rebounded in 1999 and 2000 (fiscal years 2000 and 2001). Flying J stunned industry analysts by easily surpassing the annual benchmark of a 20 percent revenue increase. The rise was 26 percent in 1999 followed by an astonishing 47 percent in 2000. This latter year boost of nearly $1.4 billion was larger than the total sales of some viable competitors. The increases were significantly affected by higher fuel prices, but gallons sold in 2000 also rose by a noteworthy 20 percent. In an industry characterized by relatively flat sales, such extraordinary gains under extremely forbidding circumstances documented that Flying J is continuing to run away from its nearest competitors.

During the decade of the '90s, the company never wavered from its long-held strategies: push volume while maintaining low expense ratios; be highly leveraged at the expense of lower profit margins; put fixed assets (travel plazas) in place as rapidly as possible; and maintain a highly favorable cash flow. Annually, net profit after taxes hovered at just under 1 percent of sales except in the three years affected by the Comdata dispute. Even then, the company never experienced a loss. At the end of the decade, the debt-to-equity ratio had increased to more than 1.5 to 1, with equity rising by nearly 200 percent and long-term debt by just under 500 percent. As in the past, the net results of these decisions were positive. Sales skyrocketed by 440 percent, assets grew nearly 400 percent, and the number of facilities tripled—a clear indication that Flying J was lapping the opposition.

The importance of travel plaza expansion (versus refining, supply, development, and wholesaling) is reflected in the sales figures. In 2000, retail sales (primarily from travel plazas) accounted for 93 percent of total revenue versus 74 percent in 1991. Sales from the Conoco–Flying J partnership rose from 77 percent of revenue in 1991 to a peak of 81 percent in 1997. In 2000, this percentage dropped to 78.4 percent resulting from Conoco's decision to participate in fewer of the new plazas as noted in chapter 11.

Driver comments:
"As a team driver, we try to stop at Flying J whenever we have the chance. . . . We have found Flying J is all around one of the best places to do this. You have some of the cleanest facilities, best selection of odds and ends, and good food no matter what state we might be in. Also, the staff is some of the friendliest we deal with no matter how their day is going. You have a first-class operation."
                                                        —Glenn Potts

---

A Look at Competitors

What makes the free enterprise system the most productive in the world is competition. In fact, the stiffer the competition the more vibrant the industry, and truckstops and travel plazas are no exception. Flying J pulled a stodgy, backward industry into modern times by providing services and facilities for truckers that are the equivalent of those other travelers experience on the nation's highways. Finally a company came to recognize drivers' unique needs by adding restful lounges, spacious restroom facilities, spotless showers, and work areas where books can be kept current with the aid of Flying J's Internet hookups and other information services.

Noting Flying J's success, in the mid-1990s the entire industry began to change as truckstop owners sought to broaden and redefine their missions. Initially they moved to modernize their chains in regions where they were strong and Flying J had only a toehold. Several followed with ambitious plans to gain a nationwide presence. In the process, many copied some of Flying J's design concepts and driver incentives. Such actions only pushed Flying J to be more aggressive in protecting and expanding its lead. Top management remains vigilant, knowing that the first sign of a company's downhill slide is when it comes to rest on its laurels.

For a more complete picture of competition within the travel plaza industry, Flying J's four major contenders deserve coverage.

TravelCenters of America (TA)

Originally known as Truckstops of America, TA was founded as a chain of six stations in 1972. Company history is dominated by a series of mergers and ownership changes. Initially it was sold to Ryder before Standard Oil Company of Ohio obtained ownership in 1984. Three years later, British Petroleum purchased Sohio, and in 1993 BP sold the chain to the Clipper Group. In 1997 this company combined with NATIONAL Auto/Truckstops, another of the old domestic truckstop chains, and a new name was adopted: TravelCenters of America or TA. At the time of the merger, NATIONAL had twice the retail outlets of TA, mostly former Union 76 stations. In 1998 TA acquired 17 Burns Bros. Travel Stops, and in 1999 another merger followed with Travel Ports of America. Ownership changed again in November of 2000 when TA was acquired for $731 million by Oak Hill Partners, a private equity group founded by Robert M. Bass.

Headquartered near Cleveland, Ohio, TA at this writing has 155 facilities (121 company-owned and 34 franchises) located in 40 states. In *Forbes* magazine's November 2001 list of the 500 biggest private companies, TA ranked 96th. Revenues in 2000 were $2.06 billion; employees numbered 10,635. TA claims annual diesel sales in excess of 1 billion gallons, and has fast pay and loyalty programs.

Although its facilities are essentially equal in number to Flying J's, as one would expect from TA's frequent mergers and ownership changes, the chain lacks a standard format, has a wide variance in services and quality, and numerous facilities are sorely in need of updating. TA has announced a $300 million modernization program to refurbish 109 of its current stops and construct additional super plazas. Many will contain features similar to the signature Flying J format. Resulting from TA's ragtag facility mix, Flying J's average sales per plaza is significantly higher, services are more consistent, and a broader use of cutting-edge technology is evident.

Pilot Corporation

Pilot Corporation was formed just prior to Flying J Inc. James Haslam II started the company in 1958 in Gate City, Virginia. His operations gained early financial strength when Marathon Oil became half owner in 1965. Later, in 1988, the Haslam family bought back Marathon's interest. Pilot opened its first travel center in 1981. In *Forbes* magazine's November 2001 list of the 500 major private companies, Pilot ranked as 68th. The company at that time had 134 travel centers located in 37 states plus 46 C-stores in Tennessee and Virginia. Pilot claims to sell nearly 8 percent of over-the-road domestic diesel fuel. Company revenues in 2000 were approximately $2.568 billion and employment was 7,710.

Throughout its history, Pilot has concentrated on long-haul truck traffic. Its facilities represent the more typical truckstop rather than a multipurpose travel plaza. As a result, Pilot outlets tend to be smaller, have fewer amenities, and feature national fast-food chains. Pilot can be extremely competitive in diesel pricing since it caters primarily to professional truck drivers and has less capital tied up in each truckstop.

On March 15, 2001, Pilot announced the forming of a joint venture with Marathon Ashland Petroleum, LLC (MAP). Once finalized, the new company—known as Pilot Travel Centers LLC—began operations on September 1, 2001. MAP brought Marathon Oil back into ownership. MAP is owned 62 percent by Marathon and 38 percent by Ashland Inc. A retail subsidiary of MAP, Speedway SuperAmerica LLC (SSA), operates the facilities involved in the merger. SSA is headquartered in Enon, Ohio, and is the nation's second largest company-owned-and-operated convenience store chain with more than 2,250 outlets in 20 states. According to press releases, the merger will extend Pilot's travel center chain to 235 locations.

Petro Stopping Centers

Petro, headquartered in El Paso, Texas, was founded in 1975 by Jack Cardwell. Starting somewhat late in fuel retailing, the company has a more consistent line of modern plazas than most others, making its facilities somewhat comparable to Flying J's. As of January 2001, the company was the operator or franchiser of 51 plazas in 30 states. Nearly half of these are franchises. (Flying J purchased four of these August 1, 2001.) Petro revenues were $983.2 million in 2000, and employment 4,186. Their restaurants are known for serving quality food. Plans call for seven locations to be opened in the next two years. Driver amenities include a Rapid Reader pump card and a driver rewards program. Volvo Trucks North America Inc. owns 29 percent of Petro. Mobil Oil is a smaller minority owner. Petro, highly leveraged like Flying J, is a worthy competitor, but lacks the coast-to-coast presence, pricing advantages, and in-house developmental technology enjoyed by Flying J.

Williams TravelCenters

Williams TravelCenters is part of Williams Energy, a $25 billion enterprise that specializes in a variety of energy products and services. In 2000, the Tulsa, Oklahoma, based company ranked 168th on the *Fortune* list of the 1,000 largest domestic corporations. That year it had $11.481 billion in revenues and profits of $524 million. The company, founded in 1908, has been known historically for its extensive pipeline ownership, both domestic and worldwide. Williams operates the nation's largest-volume system of interstate natural gas pipelines and is one of the nation's major producers and marketers of natural gas and propane. Currently it is the second-ranked power provider to the Los Angeles basin.

In its goal to become an energy superstore, several years ago Williams moved aggressively into fuel retailing by purchasing C-stores and building travel plazas. In 1998 the company merged with MAPCO, the owner of several hundred C-stores in the Tennessee metropolitan area. Two years later (similar to Flying J), the company decided to reduce its C-store exposure and sold 198 of the MAPCO stores to Delek. The explanation given was that the company wanted to "aggressively focus on further developing businesses that more clearly match its strategies," in this instance truckstops and travel plazas.

In January 1999 Williams had 29 travel centers. The company currently operates 57 outlets along the interstates, some of which are smaller fuel stops. Nineteen of the larger formats are in various stages of development. Most of the original ones are located in states surrounding Kentucky and in other areas of the South. Recently completed or under-construction plazas will extend its logo to the Southwest and Intermountain regions. Being a young division of a well-financed company, Williams's plazas are modern, well appointed, and on occasion, border on being elaborate. As a recent market entry, the company is just introducing driver incentives such as the loyalty program Flying J has had for 20 years.

In the next few years, some fallout in the travel plaza industry is likely. Annual industry revenues of $42 billion are significant, but growth is little better than inflation, even with diesel sales temporarily jumping by 6 percent in 1999. Miles driven by truckers have raised considerably, but improvements in truck aerodynamics, the technology of diesel engines, and tire design have significantly reduced fuel consumption. The current average is five miles per gallon, with new rigs cruising along at seven miles per gallon. With Flying J's annual sales increasing a minimum of 20 percent, and national consumption rising by at most 3 percent, some retailers are obviously hurting. As occurs in most industries, single owners and small chains—lacking the finances and capacity to keep up technologically—feel the pinch first. Similar to the early history of automobiles and airlines, when several hundred companies were eventually reduced to a few giants, travel plaza operators will likely experience the same washout. NATSO has 550 members owning 1,100 travel plazas, an average of two each. Two-thirds of all carriers have six trucks or less in operation. Thus, consolidation and fallout still have a long way to go. Mergers are the current trend in petroleum-related industries ranging from truck manufacturing to titans on the supply side as evidenced by the unions of Chevron and Texaco, Exxon and Mobil, Philips Petroleum with Tosco, and Conoco and Phillips. This trend is sure to slop over into the travel plaza industry, making further consolidation inevitable.

Leadership in the '90s

As noted earlier, beginning in the mid-1980s, Jay started to pull away from day-to-day operations. When Phil Adams became executive vice president in 1987, Jay continued to distance himself from the details of travel plaza development and other business undertakings. After the Conoco partnership came into being, he knew the company was secure and pointed in the right direction. Now able to implement his long-term personal goal, Jay turned over essentially all strategic and operations responsibilities to Phil, making him company president on March 26, 1991.

Jay prefers a lifestyle allowing him to "do his own thing." For several years in the '90s, he owned a 109-foot yacht that operated off the West Coast of the North American continent. He frequently sailed from Mexico to Alaska depending on the season. He generously provided numerous voyages for employees and friends, often making the total yacht available for their use. This interest has diminished, but he has never lost his attraction for flying and airplanes. He frequently buys and sells smaller jets, more out of enjoyment than to extract a large profit. With competent mechanics and extensive hanger space at the Brigham City Airport, he has the capability to refurbish multi-passenger jets, often adding features to upgrade their performance before placing them back on the market. His interest in real estate is also undying. He owns a picturesque 1,600-acre ranch in Montana and is always one to spot a prime piece of property that others have overlooked.

Jay recognizes that running a continent-wide, $4 billion business is not his forte. He remains as chairman of the board and generally attends weekly management meetings. Phil has filled in as president beyond Jay's fondest dreams, and Flying J employees are no less in awe of what the new president has accomplished. Phil's vision of the industry has no bounds. By concentrating on the customer rather than on trucks, travel plaza, and/or fuel, he has taken a holistic approach that has moved the company into

Jay and Tamra's home site at the Dancing Wind Ranch in Paradise Valley, Montana. (July 2000)

banking, insurance, credit cards, phone cards, a variety of information and communication services, shopping opportunities, and other potential attractions for truck operators and their families. With customers in the cross hairs of his strategy, Phil relied on an assortment of driver incentives and superior technology to leap ahead of competitors. Thus, during the '90s decade, Flying J continued to develop unequaled loyalty and hospitality programs and strove to put the evolving electronic marvels at the fingertips, eyes, and ears of drivers. The result has been to be the front-runner in point-of-sale, communication, and information systems that benefit both management and customers. The strategy is simple but effective: satisfy customers through unexcelled hospitality by broadly defining their needs and by taking advantage of leading-edge technology.

Phil is a master strategist who anticipates opportunities before most others have any inkling of their existence. His colleagues are constantly amazed at how he simultaneously keeps numerous balls in the air, and how he moves through every facet of the company with an overall knowledge equal to the person in charge. When progress in a particular portion of the business does not meet expectations, Phil gets personally involved, and new ways are found to accelerate growth. Phil enjoys the leeway offered by a private corporation and resists constraints that might hamper his ability to forge ahead when conditions dictate. He tends to grow impatient in dealings with other companies, especially those with layers of committees, causing issues to be hashed over for weeks at a time.

Rapid growth is not without its problems, and some debris is always left in its wake. Occasionally the company moves faster than it can maintain optimal staffing, and some sacrifices have been made in adequate management of human resources. Overall, it is a company on the go, and employees revel in helping overcome the myth that a small rural company has no chance to overtake the well-established, generously financed petroleum Goliaths who boast of their triple A credit ratings.

TRAVEL PLAZA OPERATIONS

*"That which you persist in doing becomes easier to do,
not that the nature of the thing has changed,
but that your power to do has increased."*
—Ralph Waldo Emerson

Plaza Formats

*T*o maintain consistency and be cost effective in construction, Flying J relies on three basic plaza designs, each varying in size but all with similar features. The company's signature facility is the large, full-service travel plaza that rests on a 20-acre site. Such generous real estate leaves room for ample parking and for facilities such as a motel, truck wash, and/or service center. The layout typically includes a spacious area for trucks that funnels into 12 diesel islands. A separate passageway and fill-up for recreation vehicles (RVs) is provided with easy access to propane tanks and an RV dump. Flying J's appeal to RV customers is unique in other ways. The company offers membership in an RV club. Participants are issued a Flying J RV credit card that offers special benefits such

RV's lined up at the Ehrenberg, Arizona plaza, March 2002.

Merchandise available at a Flying J Plaza.

Flying J headquarters building in Ogden, Utah. (June 2002)

Flying J headquarters building, Ogden, Utah.

Large format Flying J travel plaza in Texas, April 2002.

Fuel desk and pay station at the plaza.

C-store section at a large plaza.

C-store section at a large plaza.

Gift shop at a large plaza.

Flying J computer room.

Restaurant view at a large plaza.

Buffets in a sit-down restaurant.

Diesel islands for trucks at a large plaza.

Flying J tanker truck.

Truck being weighed on Flying J scales.

Signage at the Willard Bay Plaza. (February 2002)

Willard Bay, Utah, mid-to-small size plaza.

North Salt Lake refining and storage facilities. (February 2002)

View of Big West Oil facilities at North Salt Lake. (February 2002)

RVs at designated RV facilities, Ehrenberg, Arizona, plaza. (March 2002)

as discounts on fuel, merchandise, and services. As a result, the growth rate of Flying J's RV business exceeds that of its truck customers.

Fronting the premium facility are six islands for cars with card readers making it possible to avoid visiting pay stations. Parking, sufficient for upwards of 200 rigs, is paved and brightly lit at night. A set of scales is close by for drivers to insure that their vehicles meet local weight requirements.

Inside, the fuel desk is surrounded by an array of merchandise that gives one the feel of a well-stocked country store. Included are staple grocery items, snacks and sundries, hot and cold deli offerings, clothing, newspapers, magazines, souvenirs, rental book tapes, electronic equipment, a soda fountain, and other items travelers look to purchase. A connected wing includes a full-service restaurant with seating for 185. In recent models, the restaurant waiting area is adjoined by a gift shop for those with time to browse. Fast-food patrons have access to a food court with a minimum of two offerings (typically Mexican and Chinese) located closer to the main traffic area. A parlor or warming table featuring pizza by the slice is another option for quick service. With this variety of venues, the typical large plaza has 100 employees.

Commercial drivers are treated to an area that features a comfortable television lounge, game room, laundry facilities, and workspace where they can connect to the Internet, do their book work, and telephone in private. They enjoy clean, tiled, private-stall restrooms that are connected to 15 shower stalls accessed by pin numbers on a keypad. Drivers reserve showers by number and are given a private code for entry without needing to visit the fuel desk. A shower (normally priced at $5) is free with a minimum fuel purchase of 50 gallons. Each shower room contains a sink, toilet, two towels, and a new bar of soap. Ever popular with drivers, a plaza often issues more than 500 shower passes a day.

Driver comment:
"My husband is an over-the-road driver and buys all of his fuel at the Flying J's. During July of this

*Flying J Plaza
Shower Room.*

*Typical kiosks in a
Flying J plaza.*

> year I had the privilege of riding with him. We
> stopped all over the country and we ate and show-
> ered at your facilities. We also stopped at other big
> truck stops. Needless to say, we agreed that your
> facilities are outstanding. We were both treated
> very well at your stores and the shower facilities at
> all of your truckstops were the best ever!"
>
> —C & B Hester

The ultimate in modern technology is represented by a touch-screen kiosk where customers can send and receive faxes; copy documents; obtain mileage and routing information; search nationwide for freight on a back haul or to match loads; locate restaurants, motels, and services in the surrounding area; log on to company e-mail services; obtain disposable prepaid calling cards; and check their status on Flying J loyalty programs. Flying J's one-stop service centers for long-haul drivers have relieved much of the boredom and concern drivers have of being isolated while on the road.

Plaza structures and the layout of facilities, roads, and parking can be modified when a site has a peculiar size, unusual geological setting, or is subject to highly restrictive zoning regulations. Local codes often mandate specific exterior features that match surrounding architecture. Under such circumstances, Flying J modifies the plaza's external appearance, but the internal footprint generally remains the same. If warranted, the basic format is customized to appeal to a unique customer base such as the plaza near Tampa, Florida. Being located next to the nation's largest RV retailer, extra space and added services are provided for RV owners.

In most respects, the mid-range travel plaza format is a scaled-down version of the larger design. Parking spaces drop by one-fourth, showers number 9 rather than 15, the sit-down restaurant accommodates 140 patrons versus the normal 185, the number of fast-food stations is reduced, and diesel islands drop from 12 to 8. The restaurant serves the same menu as the larger facility, and hence the kitchen is identical, as are services for professional drivers.

Flying J's smallest plazas are called "fuel stops" because of their more limited services, especially for diesel customers. These units are designed for placement along state highways rather than as links in the interstate network. However, many are located within easy freeway access and, therefore, gain significant sales from 18-wheelers. Historically, company fuel stops have varied in design and services more than have plazas. Now Flying J is bringing greater uniformity to this type of facility through two alternative designs. The basic design is similar to an extremely large C-store with the addition of five diesel islands as a magnet for truckers. Many fuel stops have generous parking for up to 100 rigs. Inside, space for drivers is reduced to a lounge and an area for Internet access where they can update their records and correspond with others. Separate restroom facilities include three showers. Many fuel stops have a sit-down restaurant seating 125. The smaller of the two formats has three fast-food options with seating for 25.

Plaza Design and Construction

Flying J has a long history of designing and building its own facilities. Today these functions are handled by Property Development Group (PDG) under John McSweeney as president. He heads up the architect team responsible for drafting and design and is in charge of the unit that monitors and performs construction. The designers position the buildings, parking, gas and diesel islands, scales, and propane stations on the site for appropriate traffic flow. Determining traffic flow is always complex given the number and mixture of vehicles handled each day, especially when entry or exit involves busy freeway off- or on-ramps. Two-thirds of a plaza's customers are 4-wheel vehicles and one-third trucks of varying size, mostly class 8, 18-wheelers. However, an 18-wheeler's fuel tanks hold 200 gallons versus 15 to 21 for the typical 4-wheel automobile. As a result, the average purchase made by a truck driver is more than 100 gallons, ten times that of passenger car drivers. Because mixing 4-wheel and 18-wheel truck traffic is never satisfactory and trucks need ample space for maneuvering, all sites have separate entrances for each type of vehicle.

The construction arm of PDG consists of estimators, project managers, and some construction personnel. The actual construction is handled by subcontractors. The construction department is given no favoritism when competing on Flying J projects. PDG must be the low qualified bidder to be awarded the contract. Still, three-fourths of Flying J projects have been constructed under the talented direction of PDG.

> Driver comment:
> "About two days ago I stopped for a while at your facility in Matthews, Missouri on I-55. It was really nice to find a travel center so well maintained and attractive on both the inside and outside. It had to be someone with a lot of insight and concern that designed your facilities. It proves that people care that we have something nice to go to out here. Thank you!"
>
> —T. Wilson

The complexity of designing and constructing a travel plaza is such that even many experienced contractors are taken aback when reviewing drawings. Air, water, and electrical conduits must be extended to each fuel island. Fuel is stored in a half dozen or so 10,000 to 20,000 gallon underground, doubled-walled storage tanks connected to the islands by double-walled piping. In recent years, with transactions (payments, credit, fuel, etc.) being processed electronically, the number of CPUs (central processing units) at a location has jumped threefold, from 6 to 18. In addition, the mechanical systems must now provide varying temperature and humidity zones in the kitchen, fast-food areas, convenience store, restaurant, drivers' lounges, and showers. Construction costs for the larger plazas run from $6 to $8 million. Given the enormous complexity of such a facility, PDG maintains a remarkable record of being able to have a site up and running within six months after the bid is awarded.

The PDG division is an important element in the overall success of Flying J. To have Cadillac accommodations and still be the low-cost provider mandates that the facilities be constructed at minimal cost, add to operational efficiency, and attract customers. In fact, if the low bid is significantly above the estimate, construction contracts are not awarded out of concern that the facility may not be profitable given the razor-thin margin on fuel. Plans are scaled back or later rebid if local building costs decline.

Handling the Restaurant Dilemma

Of all plaza operations, food services are generally the most difficult to manage and keep profitable. Restaurants are labor intensive and turnover is high among both food preparers and servers. Spoilage and waste can create problems, and the variety of offerings keeps inventories high, not to mention the intense competition that comes from well-organized, long-standing chains. Restaurants are the one element of travel plaza operation that managers approach with the most trepidation. Many feel fortunate if their food services break even. In this constantly changing industry, careful monitoring is required to maintain consistent quality, satisfy ever-broadening customer tastes, and overcome the turbulence caused by unstable staffing. Running a restaurant is similar to investing in the stock market. It's difficult to beat the averages. Sound decisions can result in reasonable returns for those having experience and well-grounded judgment, but novices are likely to fail, and uncertainty is always a factor.

Restaurants rank behind fuel prices and location as the next most common reason why truckers stop at a particular site. Food sales are now second to fuel at the typical plaza. Although Flying J food services account for only 4 to 5 percent of travel plaza sales, they contribute more than $185 million to annual revenue, and the pace is increasing. This growth comes from Flying J placing more emphasis on food services and from cultural changes widespread in society. In 1960, 70 percent of the food consumed in

Inside of the Flying J restaurant, Winslow, Arizona.

Exterior restaurant signage, Willard Bay, Utah.

the U.S. was in the home. Today it is less than 50 percent and declining.

At the typical C-store, food services are more predictable (and profitable). Its customers, anxious to get back on the road, seek to purchase drinks and carryout food items that require no waiting. Gross profit margins are at least three times those of fuel, and food labor expenses are low because customers serve themselves. A travel plaza must satisfy these customers plus long-haul truck drivers, RV operators, and other highway travelers. Following a 300-mile stretch, long-haul drivers want a place to relax and enjoy a more extensive meal. A steady diet of sandwiches or convenience foods is not to their liking. These professionals typically account for 50 percent of a plaza's food service sales. Tourists and others traveling long distances in four-wheel vehicles also seek to stretch, relax, and satisfy their hunger before proceeding on. And, as Flying J has learned over the years, serving quality food at a reasonable price in a stylish setting can bring in repeat local customers and cause experienced travelers to look for a Flying J sign at mealtime.

In attempting to solve the restaurant dilemma, travel plaza operators (including those at Flying J) have experimented with a variety of food options: acquire or lease space to a national fast-food franchiser (a McDonald's, Burger King, Pizza Hut, Wendy's, etc.); establish their own fast-food eateries; obtain a franchise to operate a sit-down restaurant such as a Country Pride or a Denny's; or take the more risky approach of hiring chefs and running a sit-down restaurant. Similar to Flying J, 90 percent of all truckstop chains operate their own eateries, many under a franchise arrangement.

Over the years, Flying J has tried each of these alternatives. Until recently, customers could never be sure what food services would be offered at each location. Now the design of all large and medium-sized plazas includes a Flying J–managed sit-down restaurant, either a Cookery Restaurant & Buffet or the most recent format, a Country Market Restaurant & Buffet. Originally, the sit-down restaurants were entitled Thad's, but most of these have been converted to the two new formats. As noted, restaurants at large plazas seat a minimum of 185, and those medium sized accommodate 140. All sit-down facilities are spacious, attractively designed and furnished, and managed under the motto, "Serving good food with old-fashioned hospitality all across America."

Currently Flying J operates approximately 250 eateries of which more than half are fast food. In 1995, the new plaza design introduced a food court that included an Italian pizza and sandwich selection (Pepperoni's) and a Chinese food eatery (the Magic Dragon). Since then, a Pepperoni's parlor or cart has been added to most plazas, and the Magic Dragon can be found at 46 locations. Beginning in 1998, at select locations the company has experimented with three other fast-food options: a Felipe's Cantina serving Mexican food; the Hub, a diner concept offering breakfast, burgers, and shakes; and J Subs featuring sandwiches.

The larger sit-down restaurants provide the bulk of Flying J's food service income, and as one would expect, are the most difficult to manage. Answers to the following questions can tip profit scales in either direction: How much space should be devoted to buffets versus booths and tables? How frequently should menus be changed and specials offered? How extensive should the menu be given the various food items to be purchased and kept in inventory? To what extent will demand fluctuate during different seasons of the year? With plazas remaining open 24 hours, what portion of a day should a full menu be served?

With consumers being increasingly sophisticated in their tastes (including foreign cuisine), they expect something more than hamburgers, meat and potatoes, and fried chicken. A varied menu, pleasant atmosphere, and high standard of service are essential. Accordingly, Flying J's most recent signature sit-down restaurant (the Country Market Restaurant & Buffet) includes more than 85 separate entries ranging from buffalo wings and sandwiches to pasta, seafood, steaks and chops, chicken, and some ethnic foods. Several buffet bars are scattered throughout the restaurant, and a carving station is set up for dinner. The standard

menu is modified to meet regional differences. At the highly popular, all-you-care-to-eat buffets, breakfast options are served in the
morning, a lunch selection replaces breakfast from ll:00 a.m. until
4:00 p.m., and dinner items are featured from 4:00 p.m. until
10:00 p.m.

In its continuing efforts to address consumer trends and
expand its customer base, Flying J recently revised its menu to add
more tasty offerings and broaden its appeal to local clientele. The
ten-page menu has more health-conscious entries and nontradi-
tional dining choices such as ethnic foods. As an enticement for
families, Flying J introduced a special menu for children and an
eight-page kid's fun book with crayons.

With 250 restaurants to supply, the complexity of inventory-
ing such an array of food items challenges the imagination and
becomes even more difficult to comprehend when user statistics
are considered. Each week Flying J feeds approximately 450,000
people. Contracts for staple items (coffee, potatoes, and beef) are
negotiated each year. Contracts with major soft drink companies
are one to three years. The annual coffee contract is currently for
800,000 pounds. During one two-month promotion, 170,000
steaks were sold. With food constituting half of a meal's price, savings through bulk purchases are essential for chain restaurants to
be profitable.

Preparing delectable food at low cost is only one side of the
restaurant management coin. Pleasing customers through courteous, prompt service is the other. Although it ranks behind cleanliness and food quality, customer surveys consistently document
that "being treated courteously" is the most important aspect of
service. Flying J, in its goal to be the interstate hospitality leader,
takes great care in hiring and training employees to achieve this
end. The company is aware that from the moment customers
approach a restaurant until they leave, service plays a critical role
in the dining experience, especially at highway plazas where customers often lack time to consume meals leisurely.

Flying J takes special steps to please its core demographic, the
long-haul driver. Meals to stay or to go can be ordered from the

pump or fuel desk, and purchases can be made with coupons obtained through Flying J's loyalty program (points received through purchase of fuel and other items). All booths at the Cookery Restaurant & Buffet include telephones and some contain modem connections for computer hookups.

> Driver comment:
> "When I look for a campsite for the night, I look for clean restrooms, a smile from someone at the fuel desk, and a safe place to park my truck. The Flying J network has provided me this. Although I carry a lot of my own food, I always enjoy a meal away from the truck and I can't recall ever being really disappointed with the food or the service at a Flying J. It's nice to come in from the truck and be treated like a human again."
>
> —F. Barnett

After experimenting with various food service approaches, Flying J stands by its decision to operate its own restaurants as a means of obtaining the desired hospitality and profitability. The recent change in strategy is significant. Food services, once viewed as an indispensable sideline that hopefully would not detract from profits, is now expected to be a major contributor to the bottom line. At the current time the company is exploring various means of reducing costs of the staple food items it purchases. It is not quite to the point of growing its own coffee beans in South America, but knowing Phil, it could be something equally dramatic.

Travel Plaza Support: Transportation and Marketing

The fuel logistics of supplying a far-flung chain of plazas involve another set of operations that borders on the remarkable. The typical modern interstate travel plaza sells approximately 2 million gallons of fuel per month or 70,000 gallons per day, far more than the storage tanks' combined capacity of 110,000 gallons (80,000 diesel and 30,000 gasoline). Thus, a travel plaza manager continually strives to avoid an outage, a condition considered totally unacceptable in the trade.

In addition, all states have laws limiting the weight of fuel tanker trucks that supply plazas. In Utah, these restrictions confine a vehicle to 10,000 gallons. In other states, the maximum is as low as 6,000 gallons with 7,000 being the national average. Thus, to maintain an adequate fuel supply, the typical Flying J plaza must receive ten loads per day. Some plazas sell more than double the monthly average, and during peak demand, require a load every hour. To accommodate these large volumes, super plazas have tankers assigned that do nothing other than make the circuit between the site and a local refinery or storage area.

Flying J has been in the trucking business for nearly 40 years. You will recall that before Jay's father died, the pair jointly owned a tanker truck. From the beginning, by placing special emphasis on the transportation end of the business, Flying J has developed unusual expertise in the hauling of petroleum products and has made trucking one of the more profitable company lines. In the decade of the '90s, this element of the business grew in parallel with the expansion of the travel plaza network. In 1998, after Richard Peterson became vice president for fuel marketing, supply, and distribution, the decision was made to accelerate the growth of Flying J's fleet and boost transportation sales at a time when the industry as a whole was gaining ground at a snail's pace.

To implement this strategy, Flying J has been aggressive in pursuing contract hauling for others. Currently the company's fleet exceeds 220 trucks. Most transport Flying J fuel and petroleum products although some service outsiders. Several majors contract with Flying J to meet specific regional needs when their own trucking capacity is inadequate. In addition, many commercial companies have on-site service centers or fuel terminals that depend on companies such as Flying J to deliver bulk loads from local refineries or storage depots. Flying J's trucking strategy is similar to that of retailing—strive to be the qualified low-cost

Part of Flying J truck fleet.

provider and you will gain market share at the expense of others.

Given the company's jet-like rise during the past decade, arranging for an adequate, reasonably priced fuel supply remains an unending management challenge. The North Salt Lake refinery once met most company needs. However, now this facility provides less than 5 percent of requirements. The 1991 agreement with Conoco temporarily satisfied the partnership's thirst for diesel fuel, but even here, growth has outpaced supply. Today, Flying J must look beyond Conoco for in excess of half of its diesel. Ominous as this might appear, currently Flying J is in a much better position to keep its pipelines full. As the nation's largest retailer of low-sulfur diesel, the company brings strength to the bargaining table and is better poised to win concessions in both price and long-term commitments from major suppliers. Most regions of the country are dominated by different suppliers, resulting in Flying J doing business with essentially all majors. Generally, these contracts are for one year or less and are indexed to spot market prices, such as those on the Gulf Coast. And as occurs in all facets of this low profit-margin industry, a one or two cent per gallon difference can determine the fate of a retailer.

Beginning with the original Ogden truckstop in 1979, Flying J has targeted the independent operator as its primary diesel customer. Many carriers prefer to contract with owner-operator drivers who are paid by the mile rather than purchase their own trucks and hire drivers. Under the typical financial relationship, the carrier issues the owner-operator a credit card for making payments and drawing cash while on the road. Later, these charges are deducted from the driver's settlement check. Much of Flying J's growth comes from owner-operators who tend to become repeat customers, knowing that at Flying J they can obtain the lowest fuel prices at some of the finest highway venues. These same features appeal to small fleet owners who lack the volumes to obtain major price concessions from name-brand retailers.

Initially, Flying J shied away from the large-volume trucking fleets, primarily because the owners expected higher rebates than Flying J could absorb and still be the low-price retailer. In addition, with the typical big carrier, half of its fuel is dispensed from company terminals. As Flying J expanded and became the station of choice for the long-haul driver, managers of larger fleets have experienced increased pressure from their operators to develop contracts with the nation's largest diesel retailer. To take advantage of this growing demand, recently Flying J more than tripled its pool of fleet marketing representatives. Upwards of two dozen regionally assigned marketing specialists now cover every state and parts of Canada, resulting in contracts with large carriers becoming more common. The number of these contracts is sure to increase when the TCH card gains wider industry acceptance.

Lodging

With his interest in real estate, Jay was naturally attracted to the lodging business. Beginning in the mid-1970s with the motels in Carson City and Reno, Nevada, Flying J typically owned and operated various motels, generally adjacent to a cut-rate station.

Lodging was always a sideline business. Motels were built or purchased and then sold depending on the status of the real estate market and the company's need for cash such as when Thunderbird was acquired in 1980.

Jay became more serious about pursuing lodging locally after becoming a partner in several units outside of Utah with Bob Smith, a close friend in Oregon. Bob had worked with contractors and architects to develop a limited service hotel with two to three stories featuring comfortable minisuites. Under the Phoenix Inn name, these units have an attractive exterior and, based on a modular design, construction costs are low. Customers find the facilities appealing due to the roomy suites, microwave ovens, refrigerators, coffee makers, continental breakfast, and work area with telephone jacks for plugging in laptops.

The way Jay became involved in building somewhat similar units in Utah is typical of the way he conducts business. One day in Salt Lake City while waiting for his wife, Tamra, to get her hair styled, he spotted a one- to two-acre piece of property on Fifth South (a one-way, downtown street leading to I-15). The property, in an area ideal for hotels, had been overlooked by other investors who likely considered it too small and oblong in shape. Jay envisioned that if he purchased it plus several small attached pieces of property, he would have sufficient space to add a limited service hotel somewhat similar to a Phoenix Inn. After several negotiations, he became owner of 2.63 acres, sufficient for such an undertaking.

Two years earlier, his daughter, Crystal, had completed an MBA at Harvard Business School and was working on the East Coast in marketing. In October of 1993, she married her Harvard MBA classmate, Chuck Maggelet who was employed with AT&T. They were considering returning to the West, and Jay was hoping a business opportunity would come along to interest them. Not wanting to be involved in running the hotel, he offered Crystal part ownership assuming she would help build and manage it. (Unbeknownst to Jay, Phil was making an employment offer to Crystal at the same time.) Crystal, like her father, cherishes her

Crystal and Chuck Maggelet in front of the Salt Lake City Crystal Inn.

independence. She saw this as a chance to run her own business and quickly accepted. After many long hours of planning and monitoring construction, Jay and Crystal, with the occasional help of Chuck, had a four-story, 175-room inn completed, opening the doors in April 1994. The hotel was an instant success with occupancy rates exceeding 85 percent.

One month later, Jay approached Chuck about giving up his daytime job, working with Crystal, and getting into the hotel business full time. In July of 1994, Crystal and Chuck formed MacCall Management LLC, a company specializing in hotel development and management. With the exception of a small motel in Fargo, North Dakota, this company now operates all inns and motels owned by the Call family, MacCall Management, and Flying J.

Understanding the lodging ties among Flying J, the Call family, and MacCall Management is confusing because of varying ownership and separate titles for the inns. Flying J owns three

"BestRest" inns (Boise, Idaho; Frazier Park, California; and Ogden, Utah), one "Comfort Inn" in Cheyenne, Wyoming, the small "Flying J" motel in Fargo, North Dakota, and a Crystal Inn at the North East plaza in Maryland. There are eight other Crystal Inns, all owned in various shares by Jay, his children, and their businesses. In the Salt Lake Valley (in addition to the one on Fifth South), MacCall Management built the 128-room Midvalley inn in Murray and one in West Valley City with 124 units. Other Crystal Inns are in Brigham City, Utah; Logan, Utah; Denver, Colorado; Gulfport, Mississippi; and Great Falls, Montana (opened in June 2001). Similar to other Flying J loyalty incentives, the inns offer a "Freequent Sleeper" program. After ten stays, the eleventh night is free.

Beginning with the first plaza in Ogden, Utah, Flying J bought sufficient property at each new site to include a motel. For several reasons, the company has chosen (at least to this point) to delay building more. Capital has been limited, and top priority has been given to building plazas. Also, areas surrounding major freeway off- and on-ramps are typically bordered by several motels already, leading to an overcapacity in many areas, and few Flying J plazas are in metropolitan areas where construction costs are about the same (excluding property) but occupancies and lodging room rates tend to be higher. In addition, in most regions of the country (especially rural areas), the lodging business is highly seasonal, making it difficult to average reasonable profits. Nevertheless, in time it is likely that Flying J will make a major move, likely with another partner, to round out their large-format plazas by adding lodging accommodations.

J-Care Service Centers

In the early years, Flying J operated cut-rate stations with no need or space for a mechanic shop area, or grease bay. Thus, the company had little to do with vehicle maintenance, with minor exceptions. One such exception was the experiment with several stand-alone express automobile lube and oil change stations developed several years before they became popular by companies such as Quaker State, Texaco, and Valvoline. The other was a truck maintenance facility at the North Salt Lake refinery obtained as part of the Husky takeover. This facility is limited to servicing Flying J vehicles in the local area.

After Flying J got into the truckstop business in 1979, management gave more consideration to getting involved in truck maintenance but was deterred for the same reasons that slowed lodging: capital was lacking, and truck service centers (like restaurants) are difficult to make profitable, especially with only a few units. In addition, labor costs are high due to the centers being open 24 hours to accommodate truckers. Also, carriers are more likely to sign up with one large chain where they can consolidate their purchasing power, assuming the facilities are conveniently located. Two of Flying J's major competitors, Petro and TravelCenters of America, have service centers at most of their plazas, giving each a network that would take Flying J several years to match.

Although Flying J is currently behind, it holds several advantages in aggressively moving ahead. Similar to the way giant carriers handle fuel, most do their own maintenance and repair. Accordingly, Flying J's favorite customers—small carriers and independent operators—are the ones it could most likely attract. These customers already have high regard for the company, and truckers can earn Frequent Fueler points at the centers. In addition, developing a nationwide chain would benefit Flying J's growing fuel tanker fleet. With projections running as high as 500 trucks in five years, those running the fleet would be pleased to have such a chain available. However, Flying J trucking and J-Care centers (like other Flying J business units) treat each other as being separately owned. Thus, it is not mandatory that J-Care centers be the service provider, although it is likely given the cost performance Flying J has been able to achieve in its separate business endeavors.

The first Flying J plaza to boast of a maintenance center was the one in Ehrenberg, Arizona, opened in 1989. Following this, in

1993, two more welcomed customers at new plazas in Fargo, North Dakota, and Dallas, Texas. That year one was also purchased next to the original flagship truckstop in Ogden, Utah. In 1994, the list grew by three when the plazas opened in Gary and Indianapolis, Indiana, and in Frasier Park, California, had service centers. The eighth and last J-Care introduced was part of the Carlisle, Pennsylvania, plaza finished in 1995. The standard J-Care format at the time had five bays (two wash, two lube, one tire) set up exclusively for maintenance, not repairs. In 2001, six more centers came under the Flying J flag. These were part of the properties acquired by the company through the purchase of five Petro plazas and a Bar B truckstop in Wisconsin.

Flying J's stop-and-go commitment to service centers results from where they stand in the company's pecking order. Service centers are considered desirable, not essential, especially in comparison to the benefits derived through devoting similar financing to new plazas. Flying J obviously suffered no damage by being without maintenance centers at all locations during the 1990s when the company made gains far in excess of competitors. Being of low priority, when a business slowdown seems likely, new service centers are one of the first budget items sacrificed. Thus, when Flying J pulled in its expansionist horns following the Comdata dispute of 1996, building more centers was deferred until the outcome became clear.

In 1998, the J-Care division came under new management. An improved six-bay design was completed, and the company announced plans to add 70 service centers at existing plazas. Work was to be complete in five years backed by an investment of $100 million. These plans, like earlier ones, were shelved when financial uncertainty again blocked the way. This time it was the transportation recession of 2000 that forced many small carriers and independent operators out of business, followed by the nationwide economic recession of 2001.

Three recent improvements have helped to make company service centers more attractive. In 1999, Flying J modified its point-of-sale system to cover J-Care accounts. Under this system, a maintenance history is established for every truck serviced. These records can be used by owners and carriers to schedule maintenance, obtain comprehensive reports on their trucking fleets, and document maintenance history with a potential buyer when selling a rig. The other advance in 1999 was "Truck Check," a method for analyzing oil. Sixteen tests are performed in five minutes that reveal the status of a truck's engine, differential, and transmission, providing valuable guides for maintenance and replacement. The third improvement was recently announced by the company. Now the centers will sell parts and perform repairs for such items as alternators, brakes, wheel seals, batteries, starters, and other truck essentials.

The history of J-Care centers is unsettling due to frequent changes in those supervising the program and the frustration associated with unfilled plans. Nonetheless, J-Care centers remain as one of the company's future central thrusts.

SERVICES TO DRIVERS AND THE TRUCKING INDUSTRY

"We shape our tools, and then our tools shape us."
—Marshall McLuhan,
philosopher

*A*fter Phil took over, his long-run strategy for guiding the company was two-pronged: First, be the low-cost provider through using technology to keep ahead of competitors. Second, avoid viewing future markets as the expansion of fixed assets (such as travel plazas); instead concentrate on getting close to customers and exploring in depth the various means of satisfying their needs. By doing this, hospitality became the rallying flag that led to the extensive services Flying J now provides to truckers and the traveling public.

Beneficial application of technology has been the primary factor keeping Flying J at the industry forefront. By focusing on customer needs and using technology to satisfy them, the company developed a myriad of innovative applications involving optimal use of credit cards, the ROSS point-of-sale (POS) system, multipurpose electronic kiosks, sophisticated information systems, substitution of computer and machine operations for human effort, and bold ventures into complementary businesses such as banking, insurance, and group purchasing.

From the start, Phil displayed a contagious enthusiasm for automation that pervades the company today. As he stated, "Early we made a huge commitment to systems, and everyone around here will tell you it is the best thing we ever did." Chris Stanger, current manager of POS support, holds a similar view: "Our commitment to automation is the single biggest thing that has catapulted the company to where it is today." Jim Baker, head of operations for all interstate plazas, argues that Flying J's "creation and implementation of advanced technology . . . have had more impact on the industry than perhaps any other single development."

Flying J was not only the trailblazer in travel plaza development, but also the company pioneered in applying technology to improve and broaden services to the transportation industry. In the process, the company forced all those in long-haul trucking to retool their operations and educate their employees to use the myriad of dazzling products spewed out annually by electronic companies. Initially, drivers at Flying J stops learned to use card readers located at the pumps. This was followed with how to operate a touch-screen kiosk, and eventually to use computers to exchange information and correspond with others. As a result, the entire industry has been pulled into the electronic age, and computers have become the indispensable working tool of most drivers.

Phil's goal to lead the industry in hospitality and service began at a time when the modern travel plaza was just beginning to unfold. Then few looked beyond the immediate issue of updating trucking facilities to make them the equivalent of those enjoyed by the traveling public. Phil's early vision is reflected in a statement made in 1988 to a *New York Times* writer, "Our goal is to get independent truckers and trucking companies to lock into our brand name and see us a single source of information." Few people at the time understood what Phil had in mind. Ten years later, after introducing the trucking industry to everything from Internet hookups and paperless transactions to banking and insurance, he remains just as futuristic, telling another reporter, "There's no limit to what we can market to our customers." His reference here is to more than the long-haul customer. Added to the mix are their families, RV owners, commercial drivers not on the interstates, bus tour groups, and commuters—in effect, anyone who has reason to stop at a plaza.

Market Size and Conditions

Outsiders frequently ask, "What is Flying J doing by getting into businesses such as banking and insurance?" Again, one has to appreciate the unique characteristics of an industry—in this instance trucking—to develop a meaningful answer.

The movement of freight constitutes one of the country's largest industries with annual revenues exceeding $400 billion. Eighty percent of domestic freight is moved by truck, and 75 percent of the nation's communities rely on this form of transportation as their sole means of handling cargo. The 20 million vehicles used for business haul $5 trillion in freight each year.

A multifaceted market of this size makes it difficult to analyze by segment. The industry is highly decentralized with more than 500,000 licensed interstate motor carriers in a market where small companies and independent owner-operators far outnumber their larger counterparts. Of the 8 million trucks registered in the U.S., just over 3 million are in fleets of ten or more, and two-thirds of today's motor carriers operate six or fewer vehicles. Although the majority of commercial trucks are four- to six-wheel single units, most of the freight by weight is hauled by the 3 million 18-wheel workhorse diesels that pull 1 to 3 of the 10 million registered trailers. At the wheel of these rigs are 3 million licensed drivers, many of whom are owner-operators, although their numbers are difficult to determine because approximately one-third have long-term lease agreements with regulated carriers and hence are counted as a part of their fleets. Best estimates are that independents constitute 30 percent of all professional drivers, and own 300,000 long-haul transporters.

Though the industry is highly fragmented, this is not to imply that large carriers are without influence. Companies such as Roadway (10,600 tractors and 40,600 trailers), Hunt (10,600 tractors and 45,000 trailers), Schneider (14,000 tractors and 40,000 trailers), and Werner (7,500 tractors and 19,700 trailers) all have $1 to $3 billion in annual revenues and enjoy sufficient financial strength to be relatively independent. However, in this $400 billion industry, no one freight company captures even 1 percent of the sales.

Due to the way business is conducted, carriers involved in long-haul trucking face obstacles that are especially trying. The average long-haul driver is frequently behind the wheel ten hours a day (the federal maximum), chalks up 110,000 miles annually, is gone an average of eight days at a time, and feels fortunate to sleep in his or her own bed seven days out of the month. Average fleet pay of $41,000 plus benefits keeps income marginal. The independent operator puts out $100,000 for a new rig, is burdened with extremely high insurance charges, and is at the mercy of ever-fluctuating fuel prices. Driver turnover averages 100 to 150 percent per annum, and until just recently, recruiters were looking for 80,000 new professionals. The recession in late 2001 erased part of this deficit, but forecasters continue to expect at least a 49,000-driver shortfall for the next few years.

Besides turnover, the major continuing problem carriers grapple with is to control driver activities and expenses while they are on the road. When gone for two weeks, a driver needs credit or cash to purchase fuel, pay for truck maintenance, buy food, occasionally seek lodging, and meet other daily necessities. A carrier is reluctant to give an employee a large cash advance or issue the person a Visa or MasterCard where transactions are only rejected if the card is maxed out, and no restrictions are placed on what items can be purchased. Such generosity opens the door for a driver to use the card for personal expenses or to go for a spree in Las Vegas, abandon the truck, and move on to another job. The typical employee's inclination for abuse is probably no greater here than in any other business, but the potential for abuse is higher given the excessive turnover and employees being on the road and out of the reach of close supervision.

In the first 60 years of truck transportation, carriers eagerly searched for means to improve control over drivers' expenses. Before we explore this, though, a review of Flying J's groundbreaking program that endeared drivers to this upstart company is in order.

Loyalty Initiatives

In 1986 Flying J introduced a driver loyalty program that gained the overwhelming approval of truckers and ultimately became one of the most innovative and effective marketing tools in the entire petroleum industry. Jay's strategy was to establish a set of incentives that would encourage customers to keep returning similar to airline frequent flyer programs. Going back to the 1970s when S&H Green, Gold Strike, and other stamps were common on the retail scene, Flying J distributed stamps bearing their logo to those making purchases. These could then be redeemed for merchandise out of a catalog. Long after other stamps became unpopular, Flying J continued its program with considerable success. In the early 1970s, when the company switched its focus to truckstops, Jay came up with the "Frequent Fueler" program directed at the long-haul driver. Thanks to the point-of-sale system then in place, Jay's new brainchild required just 60 days to design and implement. (As will become evident, all Flying J customer programs and services—especially those in the incubation stage—feed off each other, and few could be cost effective alone.)

Under the program, when a customer slides a Frequent Fueler card through Flying J's Express Pay Cardreader, Frequent Fueler points are automatically assigned based on the gallons purchased. In addition, points are given for products and/or services obtained at J-Care centers (such as tires and lubrication). A 100-gallon purchase equals one Frequent Fueler point bearing a redemption value of $1. At the end of the month, if previous purchases exceed 500 gallons, Frequent Fueler coupons are mailed to the driver's address. In the early years, these coupons could be exchanged for merchandise from a catalog featuring Browning sporting goods such as shotguns, fishing gear, and golf equipment. Coupons were also accepted at Flying J restaurants and lodging facilities. Today the redemption options have been expanded to encompass almost anything a trucker might desire. The catalog contains more than 500 items ranging from tools, electronic equipment, jewelry, and sporting goods to clothes, snowmobiles, and vacations. To cash-pinched drivers, the program is a means of obtaining some luxuries that could be too expensive otherwise, causing them to calculate with care the points necessary to make their next redemptions.

Similar to the confusion noted with other industry statistics, meaningful Frequent Fueler enrollment figures are equally difficult to determine, primarily due to high driver turnover. Flying J has signed on more than 2 million drivers, approximately half of whom are still out on the road. Each quarter approximately 600,000 drivers are awarded Frequent Fueler points and new enrollments rise by 25 percent, causing a continual climb in the activity level. Of course, the enrollment gain is partially offset by drivers changing jobs or leaving the profession. A 24-page publication, *The Long Haul Letter®*, is mailed to members quarterly, and hundreds of redemptions are made each month.

The Frequent Fueler program sets the standard for Flying J's strategy of appealing directly to drivers. The coupons are awarded and mailed to them, not to the carriers. Small fleet owners appreciate such incentives that add to driver retention and job satisfaction. The appeal is not as strong for drivers in large fleets because, as noted, these carriers dispense half of their fuel from company terminals, and historically they have not been major Flying J customers.

The success of the Frequent Fueler program has been nothing less than stunning. Professional drivers, aware of its benefits, badger their carriers to sign on with Flying J. As one sign of the program's inroads, 70 percent of a new plaza's customers the first month are those who have done business with the company previously. Richard Peterson, vice president for fuel marketing, supply, and distribution, a person with a long history in the industry, calls it "the most successful marketing program I have ever known." Chris Stanger, who directs POS support and frequently visits plazas and talks to drivers, refers to it as "an absolutely tremendous program. Drivers just love accumulating Frequent Fueler points." As other truckstop companies enviously watched Flying J's growing number of Frequent Fueler converts, they have introduced programs of their own. However, none to date has come close to gaining similar driver acceptance.

> Driver comment:
> "I just felt I needed to write you to let you know how much I enjoy your products and services. Just recently I joined your Frequent Fueler Club and love it!"
> —Kevin C. Wagstaff

In the 1990s, Flying J extended its loyalty program to the nontrucking segment of its business. A customer at any plaza can be issued a Flying J Rewards card that allows a one-cent discount on gasoline if the card is shown to a clerk or swiped through a card reader at the time of purchase. Similar discounts are given on diesel and propane. In addition, purchases at Flying J restaurants, fast-food courts, and convenience stores earn additional fuel discounts based on the level of monthly purchases. Another card perk is the member-only specials available at C-stores and restaurants.

The Point-of-Sale System and Credit Operations

Until 1969, the trucking industry operated essentially on a cash basis unless the proprietor was willing to extend credit. The carrier's need for credit and greater control are what led Comdata in 1969 to issue a credit card that changed the entire preauthorization process. As noted, to meet the greater control required in trucking, cards were developed that expanded preauthorization through setting separate limits on items such as fuel, maintenance, food, etc. In addition, carriers could authorize establishments to give drivers cash, thus eliminating the need for large advances at the time of their departure. As an added service, the credit card company issues carriers detailed tracking and reporting on all purchases sorted by date, vehicle, driver, or on whatever other basis the carrier specifies. With ROSS, carriers can select from 21 different criteria (driver's license number, trip number, odometer reading, etc.) as a basis for validating (preauthorizing) purchases and issuing customized reports.

A problem that continually haunts small carriers and owner-operators is that, as a group, they are considered poor credit risks due to their weak balance sheets, the turbulence that often grips the industry, and high driver turnover. The owner-operator is burdened with the difficult (often impossible) task of obtaining credit to buy a $100,000 vehicle, purchase $5,000 in insurance, and build up adequate working capital to get by until payment for deliveries is received. Small carriers are further hampered because they cannot afford to develop the sophisticated information systems equal to those of large carriers or to those offered by transportation credit card companies. The mode of survival for the underfinanced segments of a market is to piggyback on industry leaders for technical advances, placing them at the disadvantages of being somewhat behind and frequently priced out. Thus, small carriers and drivers look for outside help in improving operations, gaining credit, and modernizing reporting.

Going back to Flying J's initial POS program, it was conceptualized in 1984 and first tested at the Boise plaza in early 1986. At the beginning the focus was on making the cashier's job easier by converting numerous manual transactions into electronic ones and capturing the output into more timely, accurate management reports. Because industry computer applications were still in their infancy, the opportunities to automate functions and improve reporting knew no bounds. At the time, a retailer of motor fuel could be overloaded with dozens of different payment methods, many requiring preapproval by third-party credit companies. Each transaction had to be reconciled on a cash or credit basis with hundreds of different parties. Internally, to maintain accountability by profit center, numerous sales and cost allocations were necessary, especially when a site included a C-store, several food services, a motel, and possibly a service center.

To efficiently handle these transactions and get improved reporting, Flying J first looked to purchase existing software, but found none satisfactory. The available software was devoted primarily to processing credit transactions, and Flying J management was seeking ways to reduce labor, speed up operations, and get more beneficial reporting. In the long run, rejecting the available systems turned out to be a fortunate turn of events. Jay then authorized information specialists within the company to write their own code. The result was a POS system that jumped at least two years ahead of competitors', a lead the company proudly holds to this day. By becoming strong in computer applications, the company opened doors equal to any it had walked through in the past.

The opportunities to reduce labor as a percentage of total costs were numerous and easy to identify. For a driver to complete paperwork at a fuel desk could take from 10 to 20 minutes depending on how long the person had to wait in line. Now credit card approvals are obtained in less than one second, and the entire process can be completed in under one minute. At the time, with fuel desks being staffed around the clock, one cashier position annually required expenditures of $80,000. In addition, the typical large plaza had four to five employees in the back office just to keep accounts and make payments, a process loaded with errors and especially time consuming in grinding out reconciliations.

The primary goal in developing the system was to automate every transaction from when an item was purchased through final posting and payment. As planned, they first attacked ways to eliminate written tickets and speed up preauthorizations. This was followed by building extensive data banks and formulating reports to aid in decision making. As examples, a daily report was developed for each travel plaza manager covering all on-site operations for the previous 24 hours. For fuel alone, 16 different transactions were summarized hourly. Hourly restaurant reports summarized sales by employee and the number of customers served. Ultimately, this information was used to assist in scheduling labor, evaluating efficiency, and identifying problems needing management's attention.

The system was expanded and refined for several years leading to the next big advance when in 1992 card readers were attached to fuel islands. For the first time drivers could activate their own fuel dispensers. The next year the company introduced lightning-fast, four-second (since reduced to one-second) authorizations through tying into a national frame relay network, superceding the standard telephone 20 to 30 second dial-up procedure common with other companies at the time.

Marrying the POS System with Flying J's Credit Card

During the late 1980s, Flying J issued a credit card that customers could use at any company outlet. The card (Interstate Trucking Services or ITS) came with a variety of features made possible by the company's versatile point-of-sale system. These new cards provided credit and volume discounts on diesel fuel with the added advantages of permitting carriers to "design their own reporting and control program to meet company needs." Use of the card required carriers to pay Flying J transaction fees, but these were considerably lower than the typical 2 percent that Visa and MasterCard charge retailers. Carriers found that by using the card, they could reduce fraud through receiving real-time feedback on a driver's location, date and time of purchase, type of expenditures made, and even fuel consumption since the last stop.

As noted, in 1996 Flying J expanded its ITS card industry-wide, changed the name to Transportation Clearinghouse (TCH), and became a third-party billing company. Phil took these actions for several reasons. First, he felt others could benefit from Flying

J's sophisticated POS system and the diverse features of the TCH card. In addition, he was seeking a more direct financial relationship with customers. In effect, he wanted to place the customer relationship back in the hands of truckstop operators rather than the billing companies. Finally, the goal was to let the free market influence where a customer purchased fuel rather than the economics of the method of payment. (Making purchases at a location because credit-related transaction fees are lower.)

Flying J specialists are constantly expanding and refining ROSS applications by developing similar systems for other segments of the company such as the restaurants and J-Care centers, and by improving the quality and usefulness of management reports. Flying J's POS system continues to set the company apart from its rivals, and is a key weapon in its battle to stay on top.

TCH Today

In 1997, within a year after TCH was established, it was converted into a limited liability company (LLC) with Flying J holding 75 percent ownership and Conoco 25 percent. TCH's growth has been accelerated by several developments. The most important occurred in 1997 when Flying J received approval to establish the Transportation Alliance Bank (TAB). This is another instance of different elements of the company benefiting from a symbiotic relationship by feeding off each other through the exchange of knowledge and sharing of costs. Introducing the TCH card would have been impossible without Flying J's highly sophisticated point-of-sale system, and extending TCH services became greatly enhanced when the company started a financial institution that qualified to be a member of the FDIC (Federal Deposit Insurance Corporation).

TCH prides itself on being a "third-party billing company with a multifaceted credit processing program specifically designed to meet the demands of the transportation industry." The unit's goal is to provide the industry "the most cost effective and efficient billing system." As indicated, TCH allows carriers to have total control over driver purchases and to tailor reports consistent with their needs. Changing these parameters has been simplified though TCH Application Software, a computer disk Flying J provides carriers giving them real-time control over services available through TCH. This software offers additional benefits to a carrier. Messages can be delivered to drivers, and fuel prices can be viewed on a map containing more than 7,000 fuel outlets nationwide, making it possible to select routes that minimize fuel consumption.

The hot financial issue in the transportation industry today involves transaction fees. The way third-party billing companies (such as TCH) receive income for their services is to charge for credit and for processing transactions. Using the example of Visa, assuming a driver charges $100 of fuel to the card, the retailer would typically pay Visa 2 percent of that purchase or $2. If the driver used a TCH card, the charge would be half, or $1. Transaction fees through Flying J are a function of two factors: the number of transactions and the time involved before the carrier or party given the credit makes reimbursement. Flying J charges $1 for all transactions a customer makes within one hour. After that, each transaction incurs additional charges that decline as the transaction volume increases. The second set of charges—based on the time value of money—increase in relation to the time involved in receiving payment. Thus, if a carrier has a security deposit with Flying J, the transaction charges are typically one-third less than settlements made daily. If payment by electronic transfer is not received until the seventh day, fees increase by approximately 50 percent over the one-day charge.

Recently, Comdata switched to charging fees based on a percentage of the purchase price rather than a fixed rate per purchase, thus increasing transaction costs. This has created considerable turmoil in the industry, especially for small carriers who, because of their size, have little leverage with third-party billing companies. As a result, NATSO and other industry groups have threatened to form their own credit company.

TCH has made special arrangements for carriers and others to make single money transfers in emergencies (e.g., a roadside breakdown) or in more normal situations such as driver settlements. TCH has available blank TCH checks for use by carriers and drivers. A carrier initiates paying for a purchase using a TCH check by contacting Flying J, arranging payment, and receiving an authorization number (money code) with a set limit. The person using the check fills in this code along with driver information in the spaces provided on the front. At the time of purchase (based on instructions listed on the face of the check), the vendor or financial institution must first contact Flying J to verify the authorization. These built-in security features of TCH checks cut down on abuse, and the associated charges are significantly less than typical money transfers such as those through Western Union.

In cooperation with the Transportation Alliance Bank, TCH offers customers a MasterCard debit card tied to a TAB checking account or line of credit. Through TCH, carriers and others can also make direct payroll deposits and driver settlements to those with TAB accounts. Frequent Fueler points are awarded for all purchases made using the card, not just for Flying J charges. One of the major benefits for TCH cardholders is that customer service is available on toll-free lines seven days a week, 24 hours a day.

Transportation Alliance Bank Inc. (TAB)

Flying J's surprising move into banking was influenced by several factors. First, reflecting Phil's strong customer orientation, he feels that there is "no relationship that you can have with a customer that is stronger than a financial relationship." In addition, the industry is hindered by being undercapitalized coupled with the problem that small carriers and drivers are often considered poor credit risks. Flying J management, eager to pursue Phil's desire for closer customer ties, viewed banking as an opportunity to better serve the trucking industry that suffers from financial neglect. Besides, these are Flying J's core customers, and who has a better understanding of their needs? With this familiarity, Flying J is positioned to grant credit safely in situations normal lenders would reject. In addition—by packaging credit reporting, card transactions, and banking—the company could generate economies of scale and synergies to the benefit of all parties involved.

Accordingly, in October of 1998 Flying J obtained a banking charter from the State of Utah and established TAB, a wholly owned subsidiary. At the time, business corporations in most instances were not permitted to own financial institutions other than thrifts. Seeking a different approach, Flying J set up a management team to lead the company through the many hoops necessary to obtain a particular charter they had in mind that would qualify for FDIC insurance. Utah offered a unique opportunity, being one of three states that permit a business to own an industrial loan company, abbreviated as ILC. ILC banks serve more narrow markets than the typical bank, but are similar in being full service and members of the FDIC.

Chartering the bank opened many doors for Flying J. In addition, it gave thousands of small carriers and drivers access to a financial institution that was more understanding and sympathetic. Examples of the avenues opened and associated customer benefits include:

- Flying J teamed with MasterCard to issue debit and credit cards that are accepted at millions of locations worldwide, eliminating the limitation TCH cardholders had of only using cards at fuel stops where the merchant agrees to honor it. Credit cards were established for two specific customers: a Flying J Rewards Plus Card for gasoline customers and an RV MasterCard for RV owners. In addition, a debit card (Frequent Fueler MoneyCard) was created for those with a TAB interest-bearing checking account.

- Trucking companies can set up lines of credit, maintain deposits, and have checking accounts with the same company that provides them fuel. As noted, carriers, through their checking accounts, can make direct payroll deposits to employees who have accounts with TAB.

- In addition to typical commercial loans, carriers and drivers can apply for financing to obtain tractors and trailers, pay insurance premiums, and make other major business-related purchases.
- Carriers and truckers can obtain help in solving their liquidity difficulties. After goods are delivered, they may not receive payment for 60 days (the average is 41), putting them under pressure to meet immediate expenses like fuel and payroll. Using TAB's Accounts Receivable Management Services (ARMS) program, cash can be received much more quickly. Under this program, TAB purchases or makes advances against the invoice (accounts receivable) of the party involved and issues cash (generally within 24 hours) for 80 to 90 percent of the invoice amount. After collecting payment for the invoice, TAB remits the balance to the carrier or driver less Flying J's fee of 2 to 4 percent depending on the time required to receive payment. Accounts receivable financing is highly labor-intensive. Personal contact must be made to verify invoices, arrange for payments, pursue collections, and perform credit checks on those paying the invoices.
- Transactions at Flying J ATMs are offered at a reduced rate for those with TAB accounts. Since 1994, all ATMs at Flying J locations (now more than 160) have been installed, owned, and maintained by the company. ATMs are generally one of the most profitable business lines for all financial institutions.

Traditional commercial banks do not necessarily object to niche banks such as TAB. TAB appeals to customers the typical bank is reluctant to service. Of greater concern to regular banks are organizations such as Merrill Lynch, Metropolitan Life, Allstate Insurance, and State Farm Insurance that have entered banking following the Financial Services Modernization Act of 1999. This act tore down the traditional barriers between banks and other financial institutions. The act also opened the way for such diverse institutions as BMW, Pitney-Bowes, Nordstrom, and Drexel University to start banks.

TAB is located on the second and third floors of one of Flying J's corporate headquarter buildings in Ogden, Utah. With the bank's market being national rather than local, residents in the area hardly know that it exists. There are no large signs, billboards, newspaper advertisements, drive-up windows, or other physical evidence. Even inside the building, it is difficult to distinguish the bank from ordinary offices since there are no teller lines or patrons awaiting service.

In other ways, TAB offers services similar to the typical bank such as certificates of deposits, money market demand accounts, savings accounts, and NOW (debit card) checking accounts. Recently TAB has gone online for customers. The bank holds a bright future. In less than two years, it has become profitable, and the company is continually coming up with new applications to extend its services. The major obstacle to be overcome is gaining adequate capitalization to keep up with its potential growth.

Transportation Specialists Insurance Agency (TSIA)

Based on its expertise in transportation, in 1996, Flying J uncovered another industry niche market that was ripe for exploitation: providing insurance coverage. Phil was interested in forming a subsidiary specializing in insurance for the trucking industry. Don Arrowood who had ten years experience as an insurance trucking specialist, worked with Phil to form such a plan. Some risks were involved in jumping into a new business line, but Flying J felt that it had strong cards to play: corporate management and its employees are specialists in transportation and hospitality; the company already had a friendly relationship with thousands of potential customers (carriers and drivers), many enrolled in its Frequent Fueler program; premiums could be financed by the same corporation that issues the insurance; and TSIA could piggyback on existing company media outlets (advertising boards, television displays, and kiosks at its plazas), the *Long Haul Letter*, and inserts in mailings (e.g., those delivering Frequent Fueler coupons). As an added attraction, carriers and truckers would find it convenient to do their transportation service business with a single source, especially one that is familiar with their financial status.

Flying J was attracted to this market due to these advantages and because it was sensitive to the difficulty carriers and drivers experience in obtaining and financing insurance, especially from agents with little understanding of the industry. From Jay's perspective, such a venture should not be perceived as just another opportunity to increase profits. He made it clear that if the company did not add value to the highway customer, it did not deserve to be in the business.

After Flying J bought into Don's business plan, he was hired in March of 1997 and wrote the first insurance contract in December of that year. From the start, it was apparent that Flying J did not qualify to be a captive agency, such as State Farm or Allstate, whose agents sell only company policies. In TSIA's situation, by being an independent agency (commonly called a middleman), it can offer customers insurance from several insurers, thus giving clients multiple options. TSIA first sold insurance through John Deere and Great American. Today, policies can be written with seven different insurers. In essence, Flying J's intent was to introduce somebody else's product (insurance) to its core long-haul customer based on the assumption that the company could provide better service because of its knowledge of the industry and the particular customer involved.

Initially, TSIA found it difficult to get insurance companies to sign on. TSIA had no business in hand and wanted to sell nationwide. This conflicted with how major insurers operate. Large companies have their own coast-to-coast networks comprising hundreds of agents working small regional territories. Even the prospect of a million potential buyers through the Frequent Fueler program did little to overcome their hesitancy. In time, some came to recognize the value of Flying J's customer base, especially the opportunity to make contact with potential buyers on an almost daily basis. As Don explained to insurance executives, "Flying J can put more impressions in front of customers more often and more cheaply than anyone on the planet."

Recently, truck insurance costs have been rising 15 percent per year. The primary cause is that reinsurers are increasingly hesitant to do business with agencies that specialize in trucking. Their reasoning is that when an 80,000-pound rig is in an accident, damages are extensive. Another disadvantage is that most truckers are required to have a minimum of $750,000 in liability insurance. Hence, when involved in an accident, they typically end up paying higher claims (even if indirectly at fault) in comparison with the other parties who, in general, have much smaller coverage.

TSIA also sells personal insurance to drivers and their families. TSIA reasons that it can obtain lower premiums for this group because truck drivers are more skilled at driving than the public at large and are involved in fewer accidents. In addition, TSIA markets insurance to members of Flying J's RV Club and to the general public through its Flying J Rewards program.

The policies offered by TSIA have grown to include various forms of liability, comprehensive, collision, cargo, gap, worker's compensation, health, and occupational accident insurance. The company's goal is to split its business among several insurers to avoid being exposed to one that might change its policies or discontinue truck insurance, a danger that is especially likely in takeover situations such as occurred with John Deere.

TSIA is working toward becoming profitable. With insurance costs on the rise, it is difficult for agents to tack on typical commissions without pricing some clients out of business. In addition, overcoming the resistance some insurers have for those in the transportation industry can be challenging. Nonetheless, TSIA is growing and, in time, should be a viable addition to Flying J's portfolio of services.

Transportation Optimization Network (TON)

After the initial point-of-sale system was established in the middle 1980s, Phil recognized that much more could be done to exploit evolving technology. Some of these ideas he placed on the back burner until plaza development became more secure. One feature of the industry that obviously needed attention was hauling of freight. Freight activities at the time suffered the ill effects of fragmentation. Some small segments operated efficiently, but in general, activities were disjointed, overlapping, and disconnected. Trucks would often haul loads several hundred miles only to return empty on the back haul. Prior to trucking deregulation in 1980, approximately 70 brokers nationwide were involved as third-party intermediaries between shippers and truckers, primarily those dealing in household goods. Following deregulation, the number of brokers jumped to 2,214 as they moved into all freight transactions by commercial carriers. For more than a decade, the complexity of hundreds of manufacturers, shippers, drivers, and trucks being serviced by a multitude of brokers, dispatchers, and company agents formed a confusing matrix forcing participants to do most business in small networks.

In the summer of 1992, Flying J established a subsidiary, TON Services Inc., with Joe Kelley, a highly innovative technical expert, as the head. TON opened with a much broader charter than solving freight issues. Following Jay's practice of avoiding constraints, TON was to use technology to develop information systems and automated services that would benefit the trucking industry and motoring public. However, TON's first assignment was clear: design an electronic system for brokering freight industry-wide.

Hand-in-glove with the process of attacking the freight issue, TON's employees (initially a half dozen, now 50) developed an astonishing array of automated systems available through a kiosk that pushed the company into new frontiers. As noted, from a Flying J kiosk, patrons can match loads, post freight availability, obtain a printout showing the shortest route to a destination, gather weather reports, send and receive faxes through an assigned mailbox, purchase prepaid calling cards, and copy documents. The callboard at each plaza posts local services similar to callboards at major airports. Motels, retail shops, truck and equipment dealers, and other service providers in the immediate area advertise on the kiosks, video monitors, and displays located in the driver's lounge, convenience store, and restaurants. Soon after the introduction of kiosks, TON made access easier by developing DataLink software that allows customers to dial directly into Flying J headquarters for these services without being forced to visit a plaza.

When one first approaches a kiosk, it appears forbidding. However, those willing to experiment soon find it user-friendly, as easy if not easier to operate than an ATM machine. A keyboard is positioned below the monitor, but most operations can be handled by responding to questions through prompts on a touch screen.

The load-matching service that led to creating the kiosk was named LoadPlanner. A party can post freight availability and equipment by telephone or by using Flying J's DataLink software. Posting freight involves a fee while equipment listings are free. Savings are significant, up to 75 percent for LTL (less-than-truck-load) shippers when they are able to fill up their trailers. Potential savings extend beyond equipment operating costs. Inventories and warehouse space can be reduced, and time is shortened in making contacts to optimize loads.

Flying J has greatly expanded its national clearinghouse activities with more than 50,000 new postings each day. In 1997 TON introduced a more comprehensive freight-matching service that, when fully implemented in 1999, became known as LoadDirect. Using this software, dispatchers, drivers, shippers, and transportation intermediaries can communicate anytime, anywhere by telephone, fax, or computers. LoadDirect is available over the Internet, at RoadLink kiosks, and by 24-hour telephone access to customer service representatives. With the FaxMatch feature, once someone enters available equipment or freight into the database, the party can arrange to receive faxes showing matching loads or equipment. LoadDirect prices are based on a customer's level of monthly transactions. Subscribers can pay as little as $19.95 monthly for limited usage with subscriptions climbing to $500 a month for unlimited transactions.

> Customer comment:
>
> "Thank you LoadDirect! Your customer service is great and we love your immediate response when we post or delete a load. LoadDirect meets all of our freight matching requirements."
>
> —R & R Transportation

The services available through a kiosk were expanded again in 1998: customers can obtain insurance quotes and send flowers; Frequent Fueler card members can make address changes, arrange for card replacement, and receive points for purchases; and drivers can send and receive e-mails and scan documents to be sent to third parties.

TON remains the national leader in freight matching. However, competition has grown. One set of major carriers developed its own freight clearinghouse, primarily to better compete against a few industry giants. And, as Kelley observed, "Lots of companies are doing posting on the Internet." One such company, Internet Truckstop, offers a variety of services available only on the Net such as load matching, truck routing, fuel pricing, credit verification, and reports on road conditions and weather. Regarding kiosks, no other company has been able to make them cost effective in truckstop chains. A Florida company, PNV Inc., worked with America Online to create a trucking marketplace site that offered load matching. The company developed contracts with more than half of the major truckstop chains including Pilot, Petro, Williams, and Travel Centers of America. It had 400 kiosks installed before declaring bankruptcy in 2001.

Other companies, such as British Petroleum (BP), are also experimenting with kiosks. BP is one of the world's oil superpowers with 2000 revenues of $148 billion. It owns hundreds of retail facilities, mostly smaller units for gasoline customers. Last year this division's sales of $2.6 billion ranked it behind Flying J in fuel retailing. In 2000, BP experimented with installing web links at gas pumps and adding Internet kiosks inside. Through the links at the pumps, customers can obtain local information such as traffic and weather reports. For a fee, those using the inside kiosks can read their e-mail and surf the web. BP plans to spend $200 million to make all retail facilities Internet ready and increase its current retail sales coming from nonfuel goods. With retailers suffering from ever-present low fuel margins, BP aims to increase profits by boosting nonfuel sales from the current 20 percent to 50 percent, primarily through the difficult challenge of expanding corner gas station services by adding online retail shopping opportunities.

Operating a freight clearinghouse is not as simple or lucrative as it might appear. Brokers and others tend to list the loads that are difficult to move. As Kelley observed, "Those with better loads do not want to throw them into the winds for others to take." This is probably the leading factor behind the high turnover in those attempting to copy Flying J.

In fulfilling its role of providing high-tech services for the nation's transportation industry, TON caught the next wave of electronic development—this time telecommunications. TON's first venture was phone cards followed by expanding into wireless communications, broadband services,[1] Internet applications, and being a long distance provider. Long distance is the only game in the transportation industry, comprising 90 percent of all calls. Within a short period, Flying J acquired ownership of all plaza pay phones (now 4,000 in number) and purchased the switching needed to carry the calls. This was followed by broadband hookups for trucks in Flying J parking lots and offering carriers data services, long distance products, and being an ISP (Internet service provider). Given this capability, Flying J can go to a transportation company and say, "We'll be your long distance provider," an offer that includes the company's voice, data, and Internet requirements.

Flying J's communications program is marketed through TCH Communications with TON having operational responsi-

---

1. A standard telephone line can move 54,000 bits a second; a T1 broadband circuit moves 1.54 million bits a second.

bility. TCH Communications' long distance package is relatively inexpensive, especially since it is state-of-the-art, fiber optic, and global. Rates become especially attractive when considering the integration of the various communication services involved and discount packages relating to other Flying J services.

As noted, Flying J owns the switching and buys time from carriers that run the traffic such as MCI WorldCom and Qwest. TON is a registered long distance carrier in all 48 states. It is through WorldCom's Frame Relay that TON provides customers Internet access.

Two other recent developments give insight into the direction TON is moving. In the fall of 2001, web stations were being added to all major company plazas. These stations contain units of extremely high quality that are easy to operate and have large monitors that enhance visibility. A customer can insert a dollar bill or a credit card into the unit and have high-speed broadband access to all Internet features including e-mail. Data jacks are also included at plazas. Using these jacks, a driver can plug in a laptop and obtain high-speed broadband Internet service that includes ISP access when appropriate.

The current hot industry race is to combine all telecommunication devices a driver might use into one unit that will fit in the cab, thus giving the occupant full access while in a parking lot and to some extent on the road. Miniaturization and integration have been the trends in all electronic equipment from camcorders to handheld computers. Achieving these same results by compressing all appropriate telecommunication devices into a convenient front seat location is extremely challenging. Under the registered name of CabCom, TON will roll out a unit in 2002 that duplicates all services obtainable through a kiosk plus those involving voice and data transmission. Through this cab-mounted equipment, a driver will be able to browse the Net, view movies (like those on cable TV), have voice or e-mail communication with others, update records, and perform various related functions. Two equipment options are planned: a 6.4-inch or a 10.4-inch display. A significant advantage is that when parked in a Flying J lot, a driver has broadband, wireless (no cables attached) access to the outside world.

By compressing these functions into one unit, CabCom has produced a futuristic combination of telecommunication and electronic services with a state-of-the-art system based on industry standards. This electronic box will allow splitting of operating costs between the company and the driver depending on the services each party desires and can afford. While others are attempting to give the driver cab-access to this array of services, Flying J is again leading the pack. No other company has the experience or in-house capability to keep up to Flying J in developing the instrumentation and software necessary to merge the related functions. Furthermore, having access to wireless broadband is dependent on the distance from switches, and through its plazas, Flying J has a real estate network few others can match.

Many drivers are bewildered by today's various forms of lightning-fast electronic communication compared to the CB radio that was state-of-the-art just a few years ago. Adjusting to this pace of change is a weeding-out process in all professions. Yet, most carriers and industry analysts are surprised at how quickly transportation work forces have adapted. Companies realize that staying competitive depends on being able to keep pace with these rapid developments, which puts Flying J in an enviable position. The company has the industry's leading group of technical experts who have been at the innovation forefront in creating new services, many on the perimeter of traditional truckstop operations. No one can predict what future industry course technology will take, but it is safe to say that Flying J will be one of the major contributors.

REFINING, SUPPLY, EXPLORATION, AND PRODUCTION

"Difficulty, my brethren, is the nurse of greatness."
—William Cullen Bryant

As noted in chapter 12, in 1998 Flying J converted Big West Oil Inc. into Big West Oil LLC. Buzz Germer remained as president with some changes occurring in the business units within Big West. In addition to the North Salt Lake refinery, two other components now comprise the company. The first is Big West Transportation, a unit that owns and operates 25 to 30 trucks that transport crude oil and other raw petroleum products to the refinery or to pipeline loading stations. Following refining, these trucks deliver finished products to customers primarily in Utah and southern Idaho. The third unit of Big West is the remaining Flying J C-stores, five in Utah and three in Idaho. Currently, a former Trailside outlet in Logan, Utah (previously owned by Reuel), is being refurbished that will increase the number to nine. These C-stores are not wallflowers that have been rejected by potential buyers. They are profitable, modern units that do significant business based on their sizes.

Flying J's history would be incomplete without including the refinery's vital role in supplying cash flow that significantly contributed to the development of the retail side of the company.

The National Refining Marketplace

In the decade from 1991 through 2000, refineries suffered through the same malaise that stagnated the rest of the petroleum industry. While gross domestic product was increasing at an average 3.8 percent annual rate, petroleum consumption plodded along at a 1.7 percent pace, or 16 percent for the decade. Still, refinery capacity grew only 6.5 percent during the same ten years, pushing plant utilization levels in 2000 to more than 90 percent. Even more threatening to the industry, the rate of return for refining was 4 percent, less than that on passbook savings, and one-third that of other industries. As a result, refiners during the decade had little incentive to increase capacity. Refining is one of the most cyclical submarkets within the petroleum industry. Refineries are extremely profitable when crude prices are low and demand is high, but they can reverse themselves as rapidly as a bird in flight once things move in opposite directions.

The domestic refining industry is dominated by large production facilities. At the time of deregulation in 1981, the country had 315 operating refineries. By January 1, 2001, this number had dropped to 155. The fallout consisted of smaller refineries that are inherently less efficient and weakly financed. This financial deficiency leaves them with little staying power to suffer through long droughts and inability to pay for the improvements necessary to meet the ever more stringent pollution standards contained in the Clean Air Acts of 1970 and 1990.

Most of the nation's refining capacity is on the seacoasts or inland waterways where crude oil can be delivered by tanker or barge at a low price. Of the 155 operating refineries, nearly half are in three states: 27 in Texas, 23 in California, and 17 in Louisiana.

The industry is dominated by approximately 20 gigantic proces-sors. Two Exxon Mobil refineries (in Louisiana and Texas) sepa-rately process more than 500,000 barrels per day. (A barrel is 42 gallons, giving each a daily output of an astonishing 21 million gal-lons.) The average U.S. refinery processes 110,000 barrels daily, a figure diluted by smaller inland facilities in states such as Utah, Colorado, Montana, and Wyoming. The capacity of a medium-sized Gulf Coast refinery exceeds that of all five Utah producers.

The pipeline network that distributes most of the nation's fuel has as its primary source the Gulf Coast. From there, product is fed to the South, Midwest, and East Coast. The pipeline system along the West Coast is quite negligible in comparison. The coun-try's current pipeline network dates back to World War II when construction mushroomed and larger pipes, up to 24 inches, became operable. Capacity has significantly increased since that date but more from expanding the diameter of lines rather than adding major new ones. Pipelines also dominate in transporting crude oil. More than 80 percent of all crude is delivered to refiner-ies through these continuous flow conveyances.

The large number of small inland refineries that existed from 1940 through 1980 also can be attributed to the war effort and follow-on government subsidization. With government support, many small refineries were constructed inland during the 1940s for the purpose of moving some of the nation's refining capacity away from coastal waterways where it would be subject to easy enemy attack. Until 1981, the federal government continued to subsidize small refineries to avoid overreliance on imports. Then deregulation and the termination of the small refinery bias forced most to close when large refineries easily won out in markets where both compete. Since that date, few new domestic refineries have been constructed, the last being in 1974. As the demand for refined products (mainly gasoline) leveled off, improved technol-ogy made it cheaper to upgrade existing refineries rather than start from scratch. Today a major refinery built from the ground up will cost in excess of $1 billion, making investors wary given the slow growth in demand and the cyclical nature of the industry. As a result, in 2000, one factor causing the rapid rise in fuel prices was that refineries were already operating at near full capacity and could do little to boost output.

The surviving small refineries are those in protected markets such as the five in Utah and the Sinclair refinery in western Wyoming. Pipelines were never extended into this area from the Gulf states or elsewhere, making it awkward for the giant refiners to compete. The economy-of-scale advantages these Goliaths experience is offset by the estimated 7 to 10 cents a gallon paid for transporting gasoline and other fuels 1,500 miles. This transportation differential favors local Intermountain refineries because their inputs (raw materials) and the market for their outputs (refined products) are within reasonable distance, making transportation expenses a smaller portion of total costs compared with outsiders who might intervene.

Small refineries experience two other benefits from being isolated: By relying on regional sources of crude oil and other inputs, they are somewhat shielded from international events such as OPEC nations suddenly adding or reducing supplies. In addition, major refiners along the coasts, when burdened with an oversupply, cannot marginally price their excesses and dump them on the local market, forcing resident companies to take big losses.

The Utah Market

Six refineries currently operate in the Utah market and portions of surrounding states. In addition to Flying J's Big West Oil refinery in North Salt Lake, four others are located within a few miles. These units with their associated capacities are Tesoro (58,000 b/d), Chevron (47,000 b/d), Phillips (25,000 b/d), and a small Inland refinery owned by Silver Eagle (11,000 b/d). The other regional refinery is the one owned by Sinclair near Rawlins, Wyoming, that currently has a 62,000 b/d capacity. With the possible exception of the Inland refinery, each is reasonably profitable given the area's growing demand and lack of competition from the coastal giants.

The evolution of the North Salt Lake refinery (acquired by Flying J as part of the Husky takeover in 1986) has been influenced by the various market conditions described above. Built by Western States Refinery in 1948, this tiny unit at the time was little more than a rudimentary "teakettle" designed to produce kerosene and stove oil. Gasoline was first refined in small quantities in the 1950s before a catalytic cracking unit was added in 1962 that significantly increased output.[1] The refinery changed hands several times before Husky became the owner. Husky had made significant improvements before Flying J took charge in 1986. The refinery was processing an average of 15,000 b/d with a peak potential of 25,000 barrels daily with minor upgrading. The refinery was especially attractive to Flying J because the facility was making a profit and did not immediately require large capital sums to keep in line with the competition.

To comprehend refinery economics and terminology, a simplified summary of the refining process is appropriate. Be aware that yields can vary significantly from refinery to refinery based on the characteristics of the feedstock input and the sophistication of the refining equipment. In addition, more than one cycle is involved in different phases of the processing.

The first major step is a distillation process whereby feedstock (oil in its various forms) is brought to a boil and mostly vaporized. The light (smaller-molecule) portion of crude that becomes gasoline vaporizes first due to its lower boiling temperature and then condenses at lower temperatures near the top of the distilling tower. The next heavier-molecule portion then vaporizes as the temperature increases. This vapor condenses near the middle of the tower, resulting in products such as distillate fuel. The heavy-molecule residuals are not boiled off through heating, which brings the importance of a catalytic cracker into play. Catalytic cracking is the chemical process that uses heat and a catalyst to

---

1. A catalytic cracker is a unit used in the refining process that employs high heat and a chemical catalyst to change crude oil and other inputs into gasoline and other fuels, doubling the normal output.

tear apart low-value, heavy-molecule residuals, turning most into smaller lower-boiling molecules that, when recycled through the distilling tower, add to the per barrel crude yields of gasoline and diesel. In the distillation process, separate streams of refined products (gasoline, diesel fuel, heating oil, jet fuel, and specialty wax) are stripped off at different levels of the column as each condenses.

As in retailing, technology has been the long-run key to staying competitive in refining. As one example of early efficiency gains, in 1935, 27 percent of a refinery's output was low-value residual fuel. Today it is 2 percent or less at the Big West Oil refinery. In the 1990s, 29 refineries were closed in the U.S. and yet national output increased 105,000 barrels a day. Technology resulted in per barrel refining costs declining by an average of 6 percent annually from 1985 to 1995.

Two years after acquiring the facility, Flying J added two new processing units that greatly enhanced the economics. This began an ongoing activity of periodically upgrading instrumentation and building capacity until the refinery became one of the more modern, efficient units in the region. To reduce sulfur dioxide emissions, in excess of $20 million has been spent, making it one of the cleanest refineries nationally. Under Jeff Utley, refinery manager, it is also one of the best managed.

Currently Flying J is replacing the catalytic cracker built in 1962 with a $30 million unit that is at the extreme cutting edge of refining technology. Only one other version of this cracker has been built previously. At $30 million, this upgrade is the largest single investment Flying J has made to date for any one facility, reflecting the confidence the company has in the refinery continuing to be a major component of its business and a significant money machine. The new cracking unit, scheduled to come on-line in March 2002, will provide many benefits: refining capacity will increase from a current maximum of 27,000 barrels per day to 30,000 to 35,000 barrels; higher yields of premium products—namely gasoline and diesel—will be obtained from the same quantity of feedstock (inputs); reliability will be improved; average per unit operating costs will be lower; and a wider variety of feedstock

*Installation of new
catalytic cracker at the
North Salt Lake
Refinery, March 2002.*

can be processed, especially the heavier ends of crude oil.

The North Salt Lake refinery obtains most of its raw petroleum from eastern Utah, the lower half of Wyoming, and Alberta, Canada. Crude oil provides 75 percent of the refinery's feedstock. The balance is from sources such as butane, natural gas liquids, and unfinished high-sulfur diesel sent from other North Salt Lake refineries that lack the processing capability to meet Clean Air Act standards. Big West Oil uses two types of crude as feedstock—that high in wax requiring shipment by truck, and lighter crude transported through pipelines. The heavy wax crude (called yellow and black) is found in the area surrounding Duchesne, Utah. This crude must be trucked because it is too thick without warming to flow through pipelines. Two pipeline options are available for delivering the thinner crude. One starts from Colorado and the other from southwest Wyoming. This type of crude is differentiated by two characteristics: "sweet" is low in sulfur content; "light" weighs less, is more easily refined, and has a higher output of motor fuels.

The sweet crude Big West Oil uses is delivered through British Petroleum pipelines from sources in southwest Wyoming and eastern Utah. These lines do not directly connect to Flying J wells. Big West Oil trucks load the crude from storage tanks by the wells and deliver it to pipeline downloading stations (called injection points) in locations such as Granger and Bridger, Wyoming. The light synthetic crude used by the refinery is extracted from tar sands in Alberta, Canada. Then it is transported by pipeline to a connection near Casper, Wyoming, and on into Utah for processing.

Regarding finished products (gasoline and diesel), the only pipeline that delivers these fuels to the Salt Lake Valley was built in 1953. Known as the Pioneer Pipe Line (now owned by Sinclair and Conoco), it is loaded with fuel from the Sinclair refinery near Rawlins, Wyoming, and distributed to terminals in North Salt Lake. Conoco also uses the pipeline as the final leg in transporting fuel from its refinery in Billings, Montana.

Refineries have varying capacities to produce finished products to their liking. However, as noted, crude cannot be converted entirely to gasoline or a particular fuel, with the result that some less valuable products and "dredges" are common in all processing. Depending on the sophistication of their equipment, refineries are designed to maximize the output of those products in highest demand and those with the largest profit margins. Flying J's goal is to put out as much diesel and gasoline as possible to support its local retail businesses. Over time, the processing at Flying J's refinery has been upgraded to where approximately 80 percent of the yield is these two fuels, an extremely favorable ratio. (Roughly 41 percent is gasoline and 37 percent high-grade diesel.) Two-thirds of these motor fuels are delivered by truck to customers in Utah, southern Idaho, eastern Nevada, and pieces of

Wyoming and Colorado. More than half the gasoline and 85 percent of the diesel are sold to Flying J retail outlets. That fuel not sold locally flows from Salt Lake City through a Chevron-owned, common carrier, 45,000 b/d pipeline that runs to Pasco, Washington. Big West has drop-off terminals in Pocatello, Burley, and Boise, Idaho.

Another Big West Oil refined product of importance is wax. The thick black and yellow crudes that require trucking are ideal for producing wax. Currently 12 percent of the refinery's yield is a specialty wax. This wax, still in raw form, has an unusually high quality, making it attractive to a number of different buyers, primarily in the eastern and the southern portions of the country. Two-thirds of this material is shipped by railroad tank car to customers who use it in a variety of products such as candles, chewing gum, tires, waxed paper, cosmetics, milk cartons, Chap Stick, Vaseline, lotions, and finished wax. One-third goes to companies that mix sawdust with the wax to make fire logs sold through grocery stores and other outlets. North American companies that produce fire logs depend on Big West Oil for one-third of their wax. For several years, Flying J has had plans to build a wax plant with other investors near the North Salt Lake refinery, and when capital becomes available for this expensive undertaking the company will likely move ahead.

Small quantities of various other petroleum materials are by-products of the refining process. Two of the more significant for the Big West refinery are propane gas (sold locally), and heavy fuel oils used as an energy source by power plants, steamships, and other major industrial users.

The refinery is atypical in being able to buck the catastrophic cyclical trends that often unsettle the industry. It has never experienced an unprofitable year. To the contrary, it has been one of the company's most dependable and significant sources of cash flow. For Flying J, the refinery has been the rudder that in the early years kept the company on its financial course to be the travel plaza leader along the interstates. As refinery manager Jeff Utley states, "The refinery put Flying J in another league."

The Future

What the future holds is uncertain. With population in the Intermountain region growing rapidly and no refining infrastructure to keep up to demand, the supply and demand equation is rapidly getting out of balance, especially in summer months when consumption soars. According to the Utah Office of Resource Planning, production from Utah's five refineries has increased only 1 percent per year since 1990 while local demand has grown 4.1 percent annually, double the national rate. Continuing small incremental increases in the capacities of local refineries will do little to overcome such a wide disparity.

Of course, this pending imbalance has not gone unnoticed. Sinclair and Conoco in the spring of 2001 finished replacing their eight-inch pipeline with a 12-inch line, expanding the volume coming into Utah from 48,000 barrels a day to 70,000 barrels. Williams Pipeline (part of Williams Energy, a company described in chapter 12) has an existing pipeline running from Oklahoma to New Mexico and then to Crescent Junction, 30 miles northwest of Moab, Utah. Williams has received approval to build a 260-mile long, 12-inch diameter extension from that point to North Salt Lake. This pipeline will carry 75,000 barrels per day of finished petroleum products, increasing by 50 percent that available in the local market from the five Utah refineries. The Williams pipeline is scheduled for completion in 2004 at the point when forecasters predict that current regional refining capacity will be insufficient to meet the area's needs. At this time, construction on the pipeline is being held up by lawsuits challenging the Bureau of Land Management's authorization to proceed. Whether fuel through the Williams line will be priced low enough (given the added transportation costs) to undermine local refiners is uncertain. Flying J is confident that with its modern plant, efficient operations, and favorable supply and distribution networks, it can be competitive regardless of the opposition.

Still, one of the major clouds hanging over the heads of refiners is meeting the future clean air standards issued by the

Environmental Protection Agency in late 2000. Under earlier amendments to the Clean Air Act, refiners were required to remove 90 percent of the particulate emissions from on-highway diesel. The new rules will require refiners to reduce on-road diesel from the current level of 350 parts per million to 15 PPM phased in from 2006 through 2010. The American Petroleum Institute estimates that it will require $8 billion for the industry to come into compliance and will add 11.6 cents to the price of a gallon of diesel fuel. Small refiners who are short on financing will again be hit the hardest. It is likely that some Utah refineries will shut down rather than go to the expense of trying to meet the standards. Flying J currently is within the gasoline standards and is close to matching diesel requirements, placing the company in line to ride out any further turbulence that might be disruptive.

Exploration and Production

The Flying J division engaged in the exploration, development, acquisition, and production of crude oil and natural gas reserves is Flying J Oil & Gas Inc. This division, headed by John Scales as president, is a wholly owned subsidiary of the parent company. Headquarters are in offices at the North Salt Lake refinery. Flying J Oil & Gas Inc. owns an interest in producing oil and gas wells in Montana, North Dakota, Utah, Wyoming, Colorado, and Nevada. Current activities involve ongoing exploration projects (primarily for oil) in the Williston Basin of Montana and North Dakota, and drilling programs for natural gas in the Green River Basin of southwest Wyoming. This division is important in supplying feedstock for the North Salt Lake refinery, but financially, it is relatively independent, relying on its own cash flow for expanding operations. With several exceptions, it has financed its own growth and has contributed to the parent company's financial status. Like the refinery, it is faced with the problem of being directly exposed to the frequent wild swings in raw material petroleum prices that keep the industry on constant alert.

Growth through Acquisitions

Flying J's involvement in gas and oil production goes back to the Thunderbird purchase of 1980 as summarized in chapter 7. You will recall that one of the major reasons for pursuing this takeover was to get a more secure fuel supply. At the time, the parent organization of Thunderbird, Inter-City Gas Ltd., was eager to dispose of its retail businesses and the refineries, but it intended to hold on to Thunderbird's production assets (oil and gas wells, drilling rights, etc.). However, Jay made it clear that any deal depended on Flying J acquiring all gas and oil producing properties including current leases and mineral rights. Crude prices were high at the time and petroleum properties were being sold at a premium. The situation was not the same for refineries. Of the three refineries coming under Flying J ownership, one was down at the time, and Buzz soon had to close the others after losses mounted during the small refinery slaughter following deregulation in 1981. To the company's good fortune, the gas and oil producing properties have been profitable to this day.

At the time of the takeover, Thunderbird had oil and gas properties in Montana and North Dakota. Some assets were immediately sold off to help reduce Flying J's debt, but the gas and oil producing properties remained intact. The acquired properties consisted of part ownership in 60 oil and gas wells and extensive lease and drilling rights spread between the Williston Basin of Montana and North Dakota and the Sweetgrass Arch area near Cutbank, Montana. Production averaged 313 barrels a day. For the reader to comprehend Flying J's ownership of specific production assets is difficult because wells normally have multiple owners as a means of sharing risks and operating costs. In addition, the quantity of undeveloped leased acreage is deceptive because its quality is based on the likelihood of an area being productive rather than total acreage under lease.

The need for risk sharing results from drilling being expensive and the payoffs difficult to predict. In 1980 the ratio of dry to productive wells was eight or ten to one, especially for exploratory

Flying J well at the Williston Basin.

wells, although recent technological improvements such as horizontal drilling and three-dimensional seismic data collection have helped improve the odds. Flying J relies primarily on seismic technology. Using this approach, an energy source is set off at the surface and sound waves are measured by depth based on the travel time of these waves as they bounce off different rock formations under the earth's surface. These data are fed into computers and then massaged to create three-dimensional models containing large lateral cross sections of formations at depths two to three miles under the surface. If the results reveal a favorable structural layout and positive reservoir characteristics, drilling is initiated. In the Rocky Mountain region, a typical well will have a depth of 5,000 to 12,000 feet. On occasion, Flying J has drilled to 17,000 feet in search of these illusive minerals. Average drilling costs are in the neighborhood of $1 million with a low of $150,000 to a high in excess of $3 million. (Questar, a local utility, once invested $17 million in a single well.)

This improved technology has cut exploration costs, reducing them by 5 percent per year in the 1980s and '90s. Exxon Mobil claims that such technologies have slashed its exploration costs by 85 percent in ten years. Nevertheless, drilling remains risky given the odds of being successful and the expenditures involved in each exploratory well.

Improvements have also been made in capturing oil from existing wells. Now, on average, just 35 percent of the crude located within reach of a well is pumped to the surface, primarily due to the loss of pressure that naturally occurs as fluids are extracted.[2] Recent techniques give promise of raising this average to more than 50 percent. In the Uinta Basin and other Rocky Mountain locations where Flying J acquired oil properties, oil operators feel fortunate to bring 20 percent to the surface given the difficulty in extracting the hydrocarbons from tightly compacted sandstone, limestone, and other rock types two to three miles deep in the earth.

Getting back to the area where the initial Thunderbird properties were located, currently 37 percent of Flying J's oil and gas production and 35 percent of company reserves come from the Williston Basin of eastern Montana and western North Dakota. The crude from this region is not transported to the North Salt Lake refinery. It is sold locally, or as most often occurs, product exchanges are made whereby Flying J's refinery obtains crude from a local source and another refinery (such as one in the Williston Basin) receives an offsetting amount.

The Great Western Resources Acquisition

In the 1980s from their earlier headquarters in Billings, Montana, John Scales and some of his small crew of 14 were constantly on

---

2. Using an average can be deceptive. The amount varies from reservoir to reservoir and is especially low in the Rocky Mountain region as noted.

the lookout for properties in eastern Utah that would be close to the North Salt Lake refinery. One day early in 1989 while visiting a small, company-owned field in the Uinta Basin of eastern Utah, John, along with Jim Wilson, the division's vice president of operations, came across the field office of a Houston, Texas, company— Great Western Resources (GWR). The office was in a desirable location and it appeared operations were ongoing and prosperous. When back in Montana, John called GWR's Houston headquarters and inquired if they had properties for sale. John was given little encouragement, but a few months later he received a call from GWR stating that all company properties in eastern Utah were to be sold. GWR's reserves were the yellow and black wax crudes, difficult to transport but offering some of the highest yields, especially that from yellow wax. This crude's pour point is such that, when spilled on the ground, it can be cleaned up with a shovel and must be heated to 110 degrees Fahrenheit before it becomes liquid.

John jumped at the opportunity, went to the corporate office for financial support, and through the bidding process, gained what became an extremely valuable acquisition. The purchase gave Flying J Oil & Gas an interest in 222 producing wells and more than tripled its crude production. Known reserves were impressive at 8 million net equivalent barrels.[3] The company's current crude production from the basin is approximately 1,400 barrels a day. Many Flying J executives were relieved to obtain a local source of quality crude for the refinery.

Following this acquisition John and most of his staff (increased by 23 former GWR employees) moved to new offices at the North Salt Lake refinery or to field offices in eastern Utah. Through gaining these properties, his division enjoyed a new status within the company. The purchase placed Flying J solidly in the oil production business, and it gave the company some of the direct upstream reliability it had dreamed of for many years.

---

3. "Equivalent barrels" is a term used for reporting purposes only. To combine quantities of crude oil and natural gas into one figure, natural gas volumes are converted to equivalent barrels of crude.

Adding Properties to a Growing Inventory

At the time GWR properties were being absorbed, Flying J executives were convinced that the most rapid and efficient way to expand crude and natural gas production was to purchase operating wells. Such an opportunity came along in the fall of 1995 when Cenex,[4] an agricultural cooperative based in St. Paul, Minnesota, placed its producing properties up for sale. In exchange for $40 million, Flying J increased company crude production 165 percent (from 53,000 to 130,000 barrels a month) and greatly expanded the possibility of boosting natural gas output. Flying J was especially attracted to Cenex because the acquired properties and mineral rights were mostly located in the three-state region where Flying J was knowledgeable and had current operations—the Williston Basin of North Dakota and Montana, and portions of Wyoming.

Eighteen months later, Flying J Oil & Gas made another acquisition, this time from Equitable Resources Energy Company of Houston, Texas. This purchase involved some producing wells in the Powder River Basin of Wyoming and undeveloped acreage in various Rocky Mountain basins. The properties boosted Flying J's oil and gas production by nearly 40 percent, escalating from 1.6 million to 2.2 million net equivalent barrels per year. From the Powder River Basin, Flying J Oil & Gas obtains 26 percent of its production and the area holds 22 percent of company reserves.

In July of 1997, four months after the Equitable acquisition, the company picked up 160,000 acres of oil and gas leases in the Green River Basin of southwest Wyoming, a region known for its natural gas production. This smaller purchase, acquired at an attractive price of under $1 million, was from Apache Corporation of Houston, Texas. At this point, Flying J Oil & Gas was intent on doing more drilling, and most Rocky Mountain exploration companies were experiencing higher success with natural gas.

---

4. Cenex is an abbreviation using the first letters from the last two words in the cooperative's formal title, Farmer's Union Central Exchange.

Assuming that in the future natural gas would have greater demand, Flying J moved in this direction.

The Green River acquisition signaled Flying J Oil & Gas's revised strategy. As local oil production decreased and gas production gained in interest, it was natural for the company to switch its priorities. The new strategy was two-dimensional: keep acquiring properties that contain oil and gas reserves, and use drilling dollars to search for natural gas.

For Flying J, buying natural gas properties is often more complicated than purchasing oil wells because gas companies typically own and operate pipelines and have other related ways of gaining income beyond selling gas at the wellhead. On the other hand, gas at the wellhead is less expensive to extract and transport. In the case of Flying J, operation of 90 percent of its gas wells involves little more than turning on the valves and tying in to existing pipelines. In comparison, crude oil requires pumping, often at deep levels in the Rocky Mountain area. This pushes up operating expenses, equipment maintenance costs, and the need for above ground storage capacity. In addition, water must be separated from the crude and disposed of in other locations, adding to the labor intensity of the entire process. Another advantage natural gas has over crude is that (with the exception of 2000 and 2001) gas prices have historically been more stable. As noted, world crude prices are subject to uncertainty due to significant supply decisions being made by governments, many located halfway around the world. In North America, natural gas is largely a continental rather than a world market. Although this industry remains partially regulated, continental economic supply and demand forces generally determine pricing.

The company's current natural gas drilling activities are in the Green River Basin of Wyoming and the Piceance Basin of northwest Colorado. As of this writing, 65 percent of total company production revenue is from oil and 35 percent from gas, with the gas portion likely to increase in the future. Crude accounts for 76 percent of the company's reserves. Approximately 80 percent of the crude is first sold to the North Salt Lake refinery under the arms-length relationship described previously. However, through the trading process, only 20 percent or so ends up as direct feedstock. Natural gas is sold to processors or brokers who, in turn, sell it to major utilities such as Questar Corporation, the largest gas retailer in Utah, southwest Wyoming, and southeast Idaho.

Like other Flying J units, the oil and gas subsidiary under John Scales has experienced significant growth. The company has an interest in 965 jointly owned wells located in six Rocky Mountain states. Flying J's share of the production from these wells amounts to 6,000 equivalent barrels per day. Flying J operates approximately 300 of the jointly owned wells. This involves employees making daily inspections at each site, doing repairs, and developing recommendations on how the wells can be managed for improved performance. The company has 600,000 net leasehold acres in six Rocky Mountain states. The Flying J Oil & Gas work force has grown from 9 to 52 employees, approximately half of whom are in field offices. In addition, the company employs 20 or so contract workers in field locations. In 21 years, revenues have grown from $2.8 million in 1981 to $58.4 million in 2001.

John accurately summarized executive life in this highly cyclical industry: "You manage in good times by getting ready for the bad times, and in bad times preparing to take advantage of the good times." This volatility has been a death trap for small producers. Larger companies have attempted to escape this fate by merging to gain the size and stability necessary to protect themselves from extended U-shaped troughs. Such mergers provide some buying opportunities for smaller operators such as Flying J. Once merged, the surviving company will typically assess its recently combined property inventory and sell off non-strategic pieces to simplify management, lower costs, and reduce debt.

Being successful in the industry requires being content until opportunities show up, forcing executives to have a long-term perspective. In several instances, Flying J has come close to making merger agreements with other production and/or refining companies only to have market conditions change or other factors interfere. As R. D. Cash, Questar's chairman and CEO, stated, "Oil and gas assets never come at the time you really want them to. You've got to have patience."

As with all petroleum companies (whether they are in production, refining, or retailing), consolidation is the trend. Thirty years ago, the gas and oil industry's leading publication—*Oil and Gas Journal*—printed an annual list of the largest 500 domestic public gas and oil corporations. In recent years, the journal has reduced this ranking to 200 consistent with the remaining number of significant producers. Today there are not enough qualifying companies to reach 200. Being a private corporation, Flying J is not included. However, based on its holdings, the company would be ranked as 59th in the nation in liquids production and 60th in liquids reserves.

Flying J's Oil & Gas subsidiary has bucked sluggish industry trends as have other units of the company. This division has continued to grow and prosper while the industry has stagnated and, in some instances, declined. In the example of Utah (an accurate yardstick for the Rocky Mountain region and a state ranking 12th in the nation for oil and gas production), during the past decade—consistent with national trends—output for both products has been declining. Oil production in Utah reached its peak in 1985 at 41.029 million barrels a year. From 1991 to 2000, production dropped by 40 percent (25.9 to 15.634 million barrels). Natural gas production did not peak until 1994 at 348.14 million cubic feet. The decline for gas production during the 1991–2000 decade was much less pronounced, falling 15 percent. In general, these reductions did not result from diminishing reserves but from low prices, making marginal wells uneconomic and drilling expensive based on anticipated returns.

Growth is expected to turn around, but at a slow rate. Projections are that natural gas production and consumption nationally will outpace that for oil during the next 20 years, partially due to natural gas replacing oil in home heating and producing electricity. Flying J's Oil & Gas division is poised to take advantage of the situation with its holdings in Wyoming and Utah where the two states account for 10.4 percent of the country's natural gas reserves.

FLYING J:
THE PAST, THE PRESENT,
AND THE FUTURE

*"It's like Columbus. You set out for India. You don't find India,
but you find something else that's really cool."*
—Leonard Adleman

The Past

How Jay Call started a corporation in 1968 with assets of less than $50,000 that grew from four small gasoline stations into a company that became the national leader in interstate travel plazas with 2001 sales of $4.3 billion and assets of $1.2 billion is a story with many facets. As noted, the odds of his getting to the top were slim among the approximately 5 million U.S. corporations, 1.9 million partnerships and LLCs, and 24 million businesses of every variety, especially when most startups eventually fail.[1]

---

1. Figures accumulated in 2000 by the Internal Revenue Service based on business tax returns.

When Jay opened his initial station, he was intent on proving he was capable of being on his own, backed by a desire to create a significant business that would provide wealth and adventure. With little to start on other than his father and uncle being role models, his energy and determination brought him early success. Being young, he was extremely daring, oblivious to obstacles that would stop others, and unusually quick to learn from new experiences. He approached each day by looking for immediate opportunities rather than concentrating on what might happen ten years down the road. He soon developed an "expand or expire" mentality that rapidly spread to associates. Before long, his impatience with enlarging the business one station at a time led to the Thunderbird and Husky acquisitions that to this day confound analysts. In each case, he took over companies in situations where Flying J's equity was one-sixth of the asking price. Already being highly leveraged and deeply in debt, only his personality and skill as an entrepreneur could substitute for the normal collateral lenders would expect. Throughout his career, what many would perceive as reckless will to get ahead was tempered if not overridden by a keen sensitivity to people, markets, and circumstances.

Learning from the first truckstop in Ogden, Utah, years earlier, Jay and Phil realized that the truckstop industry was way behind the times, and that this undeveloped market gave promise of turning the company into a retail powerhouse. They had the facilities in mind and the vision to move ahead. What they woefully lacked was financing. After the somewhat troublesome experiment with franchising in the late 1980s, financing still stood in the way of developing a continental chain of travel plazas. This awakened a renewed search for a financial heavyweight the company could draw on, resulting in the 1991 partnership agreement with Conoco, one of the industry's petroleum giants. The agreement provided Flying J with the missing link in its full-fledged battle to outdo competitors. In addition, the agreement preserved the operational control Jay and Phil were unwilling to relinquish, placing Conoco in the role of being primarily a financial partner. Within a decade, Flying J was to become the country's top retailer of on-highway diesel and the front-runner in setting industry standards for modern travel plazas.

Many will say that Jay was lucky in his timing since he started when cut-rate stations were just making inroads in California and the Northwest, and he later bumped into the truckstop business in Ogden by chance. Yet, hundreds of others in small and large petroleum companies had similar opportunities but none succeeded to the same degree. Of most importance, none perceived how a chain of upscale, modern truckstops and travel plazas—seething with hospitality—could be turned into a billion-dollar enterprise serving customers on the recently completed interstate highway network.

Ultimately, Jay's greatest skill as a businessman was to recognize that what an entrepreneur seeks in his personal and professional life is often inconsistent with managing a large business enterprise. He is a true entrepreneur in that he is only content when engaged in a new project and quickly becomes bored once it stabilizes and becomes routine. In this sense, he is a pure builder, not an administrator. Rather than being the visionary who has an end state in mind, Jay, like most other entrepreneurs, is drawn in the direction that the magnet of new opportunities pulls him. Planning is involved, but it takes place in a dynamic context, much like a gambler evaluating his odds and predetermining the next step if he wins or loses in a throw of the dice. Jay is rarely without well-considered alternatives that he updates as events unfold. As an entrepreneur, he demonstrates that leadership is more an attitude and a capacity to respond rather than an ability to define precisely the future.

So the trait that sets Jay apart from others is his willingness to let go of power once he has confidence in those running day-to-day operations. Attaining power is more important for most executives than the associated monetary rewards, and few who gain such high status are willing to risk it by letting others share the stage. Yet coordinating multiple diverse units, sitting in long meetings, and paying attention to details are not Jay's forte. Ultimately, what many would consider his greatest risk and yet wisest decision was to select Phil Adams as president, a young executive with limited experience who in most corporations would have been forced to wait at least ten more years to gain such status.

Most of Jay's personal traits and his role in developing Flying J are obvious from prior chapters, but one other indicator of his influence deserves further consideration. In his case, what is clearly documented is how an enterprise takes on the characteristics of its founder, especially in the company's formative years. The philosopher Carlyle argued that every institution is the lengthened shadow of a great man. Obviously, such a statement has its limitations, but Jay's qualities and philosophy, at least in Flying J's first 20 years, can be identified in everything from the design of plazas to how employees treat customers.

Several features of Flying J's corporate culture have been noted that are a clear reflection of its leader:

- Employees are confident, not arrogant.
- Honesty and fairness are emphasized.
- Cleanliness is treated as a virtue that must be adhered to in all physical facilities.
- Privacy is important, resulting in bathroom stalls and telephone booths being separate and closed in.
- Chest thumping is avoided and publicity limited.
- Employees take risks and explore new ideas.
- Supervisors generously delegate authority and give subordinates room to grow.
- The organization is kept lean and free of unnecessary policies or rules.
- Extended, lengthy studies are avoided and decisions made expeditiously.
- Patience is valued in waiting for opportunities to develop.
- A "go for it" mentality is perpetuated.
- Resources are used judiciously, and business conducted in the least expensive, most efficient way.

One more of Jay's traits deserves further elaboration—that of being straightforward and fair. Jay realizes that if others considered someone dishonest, everything else about the person becomes suspect. Lack of integrity is a fatal flaw to someone engaged in business because it destroys the possibility of the person developing trust relationships. As he stated in a 1991 speech to Soda Springs High School graduates, "You will find that your integrity and reputation will do more for you than all of your wealth." Many examples dot his career:

- In 1963 when he was struggling financially, a former associate was experiencing losses in a station leased from Jay. Jay split the deficits with him even though it was not required under their agreement.
- An associate recalls an instance in 1977 when a potential buyer made an offer on a piece of property Flying J had for sale. Jay thought the offer was too generous and told the individual to go home and do his homework before coming back with a revised bid. The associate asked Jay why he did this. His response was, "Both of us will realize we made a mistake, and $10,000 is not enough to risk one of the principles that I live by."
- Doug Wells recalls an incident after he was first hired in wholesale sales. Jay told him, "You will be dealing in my behalf. My name is on this company. I want any deal to be good for the party you are dealing with as well as good for Flying J."
- Bruce Christensen, former vice president of Box Elder County Bank, the first financial institution to make Jay a loan, commented, "While we always attempted to get him the best rates, he knew that we also had to make some money along the line. If he made a little and we made a little, we would both be around to enjoy each others company."
- In a more recent example, Don Arrowood (the person who helped establish the recently formed TSIA insurance

wing of the company) admires what he referred to as "Jay's ethical pursuit of profit." He recalls Jay stating, "Our business all boils down to one thing: adding value to the highway traveler and a trucker's life. If we don't add value, we don't deserve to be on the list of service providers, even in hard times."

Jay becomes annoyed when others give him credit for where the company is today. He always points to Phil and then to Buzz Germer, Barre Burgon, Paul Brown, Jim Baker, Richard Peterson, John Scales, Joe Kelley, Virginia Parker, Jeff Utley, and others who have played important roles. Jay has also been fortunate in having a wife (Tamra) who is extremely supportive. Jay is correct in that only in the formative years was Flying J a one-man show. However, it is his skill in selecting the right executives and in creating an environment that allows them to maximize their personal development that moved the company to where it is today. Unlike what occurs in most large corporations, he saw the dangers of wrapping subordinates in a bureaucratic cocoon. When he gives someone the ball, he expects the person to run with it. With more than 11,500 employees, the company is fortunate to be loaded with a management team and supporting work force that are eager to push on to new frontiers in hospitality, technology, and novel ways of satisfying customers. As with most companies, the strength of its human assets, not its physical assets, is what makes Flying J great.

The Present

The economy took its worst beating in ten years in 2001. The economic slowdown of 2000 turned into a recession in 2001 exacerbated by the terrorist attacks of September 11. Real gross domestic product dipped into negative territory for the first time in a decade, average corporate profits dropped by 13 percent, and unemployment rose from 4 to 6 percent. As is customary, the petroleum market was rocked by even greater volatility. Early in the year, high oil and natural gas prices and concerns for energy shortages soon gave way to travelers staying at home and purchasing less (especially following the September terrorist attacks), causing a quick reversal in demand and an oversupply of fuel.

For the year, gasoline and diesel consumption rose slightly, but crude and refined fuel prices plummeted, causing huge declines in sales and profits for petroleum companies, especially those in retailing. In *Business Week*'s 2001 annual summary for the "fuel industry," average company sales fell 9.8 percent and profits 5 percent. Crude prices dropped from an average of $30.30 a barrel in 2000 to $20.50 in the fourth quarter of 2001. What is even more important for petroleum retailers such as Flying J is that gasoline and on-highway diesel prices both dropped 25 percent, most of this occurring in the latter part of the year following September 11. With truckstop operators (including Flying J) depending on fuel for more than 85 percent of plaza sales, when demand leveled and prices declined 25 percent, the consequences were traumatic. Losses suffered by many chains put them in jeopardy of defaulting on loans, especially since most were in the process of major expansions and thus burdened with significant debt payments. Without exception, all operators delayed construction schedules or put new construction totally on hold.

In a February 2001 speech seven months before the terrorists' attack, industry spokesman James A. Cardwell, Jr.,[2] gave forewarning of some of the difficulties ahead:

> In recent years, we have continued to feed the oversupply of fueling facilities in this country by building new sites. This forces us to keep lowering our margins—to even sell below cost at times—to maintain volumes. We simply hold our breath and hope to squeeze a living out of restaurant sales and other services. . . . Many who refinanced their

---

2. At the time, Cardwell was senior vice president of Petro Stopping Centers and chairman of NATSO, the leading truckstop industry association.

properties when times were prosperous and credit
was widely available now cannot make their payments. We failed to plan ahead and recognize what
the future holds for us.

Later in the year, the whipsaw effect of the price changes cut
into Flying J's 2002 sales. The high prices in calendar year 2000
were partially responsible for the company's 46 percent increase in
2001 sales (fiscal year ending January 31, 2001); likewise, the 25
percent fuel price declines of 2001 kept Flying J 2002 sales
(through January 31, 2002) just below the prior year $4.3 billion
level—disappointing, but a commendable achievement given the
much larger declines experienced by most of its competitors.
Profits were up 43 percent over those of 2000, but still less than 1
percent on sales after taxes. Net assets grew by more than 20 percent related in part to the 16 stations added in 2001. Nine were
newly constructed Flying J plazas—three in the Southwest, three
in the Midwest, and one each in California, Florida, and Canada.
The other seven were purchased truckstops—five former Petro
franchisees plus one each from single owners in Wisconsin and
Utah. With 157 modern plazas along the interstate system, Flying
J during the year bolstered its position as the leading seller of on-
highway diesel and the fastest growing company offering services
to truckers and travelers along interstate highways. With total gallons sold increasing in 2001 when petroleum consumption was
essentially flat, Flying J was obviously continuing to take sales
from competitors. The outcome was the familiar story in a small
margin industry: the ultimate test of a company's management
and a firm's capacity to compete is being able to survive during
extended periods of unfavorable market conditions.

One major development occurred during the year that
strengthened Flying J's financial position and resolved some of its
remaining fuel uncertainty. In October, Equilon Enterprises LLC
and Motiva Enterprises LLC, companies that at the time refined
and marketed Shell- and Texaco-branded gasolines, signed an
agreement with Flying J to co-brand a series of travel plazas. As

this agreement was being finalized, Chevron was in the process of acquiring Texaco. The U.S. government approved the merger based on Texaco's concession to divest its stakes in the Equilon and Motiva joint ventures, leaving Shell as Flying J's partner in Equilon and Shell and Saudi Aramco as joint partners with Flying J in Motiva.

The relationship under the partnership is similar to the agreement Flying J has had for more than ten years with Conoco. Flying J will operate the plazas and Shell will be primarily a financial partner. Flying J's signage and graphics will appear over the diesel islands and Shell's signage and graphics over the gasoline lanes. Shell ended up putting approximately $23 million into the partnership and, in return, five current plazas were co-branded by adding Shell's name. Conoco still has the first rights of refusal on any proposed plazas, but its participation has been less frequent as noted earlier. As of this date, besides the five plazas, Shell has opted to participate in four of the new sites to be developed in 2002. Shell's announcement of the partnership states that the new sites "will feature an increased emphasis on gasoline and RV customers."

During 2001, Flying J introduced or had under development several new ventures that gave promise of expanding its range of services and increasing company sales. The most recent edition to the Flying J family was undertaken in August of 2001 with the opening of an online store similar to the many that already exist on the Internet. Known as "E-Store," the Internet address is flyingjestore.com. This new enterprise hopes to gain sales through two Flying J strengths: First, many of the items offered are those with special appeal to truckers such as CB radio accessories, exterior accent lights, and cab accessories (bedding, privacy curtains, logbook holders, steering wheel covers, seat cushions, thermometers, clocks, seat organizers, etc.). Second, it intends to rely on Flying J's time-proven approach to doing business—offer low prices, convenience, and fast, friendly service. E-Store started with electronic products, mostly for the home such as DVD players, radios, TVs, entertainment centers, radar detectors, etc. Later, driver products, gifts, clothing, audio books, luggage, and other items were added.

Two other new products, announced in 2001 but not scheduled to be fully operational until 2002, were CabCom and PDCA. CabCom is the unit that places a powerful electronic communication center within the cab of a truck as described in chapter 14. Flying J claims that this device "will revolutionize the way the transportation industry uses mobile communication." The product—now scheduled for rollout in mid-2002—is not only convenient due to its cab location, but it greatly increases a driver's communication options and will help overcome the boredom of being on the road.

PDCA (Professional Driver Carrier Association) is a repackaging of existing Flying J services for drivers and carriers with the added benefit of obtaining savings from purchases made outside the company. The first objective of PDCA is to reward carriers and professional drivers by placing at their fingertips many of Flying J's products and services at a discounted rate. The second objective is to create an association that through its buying power will be able to obtain goods and services from others at a discount. To qualify, a driver must open a "Frequent Fueler MoneyCard MasterCard" account with Flying J's bank (TAB). As noted, this card is a MasterCard debit card that is accepted at millions of locations worldwide. Participating carriers must have a Flying J TCH (Transportation Clearing House) account and be issued a new PDCA TCH card. No special fees or dues are associated with either membership.

When PDCA members use their cards at Flying J facilities, an electronic record is kept of all fuel, food, and C-store purchases. When monthly minimum purchase levels are met, members receive rebates in their TAB accounts or credit is made to their TCH accounts. In addition, members receive special discounts for use of scales, the multiple services provided through a kiosk, purchase of insurance, J-Care center service, and so forth. Through the association's massive buying power, members can receive discounts for brand-name tires, lodging at a variety of chains, new trucks (tractors and trailers), and many other PDCA affiliate products and services. Full details regarding PDCA are available on its web page at www.pdcassoc.com.

The Future

The statement is often made that anyone who attempts to predict the future of petroleum products is a fool. In an industry where the supply of the prime product—crude oil—is determined more by political than economic forces, anything can happen. One industry analyst compared it to "dancing with the devil." Forecasts of petroleum reserves have little more credibility. Cynics like to point to the 1972 projection by the prestigious Club of Rome. Their members reached the conclusion that the world's known reserves would run out in 20 to 31 years. Many forecasters in the 1980s were predicting that crude prices would reach $90 a barrel in the coming decade when the price turned out to be $20. Technology has been the culprit responsible for many of these blunders. Improved methods of exploration led to new discoveries, better extraction techniques extended the life of existing wells, and refining efficiencies resulted in lower prices.

Organizations responsible for making projections, such as the U.S. Energy Information Administration (EIA), wisely refuse to get involved in political forecasting. Due to the large number of unknowns, the tendency of EIA and other forecasters is to make straight-line projections knowing that volatility in the industry is inevitable but unpredictable. Following are some of the latest figures from these studies covering the years 2001 through 2020:

- U.S. energy demand and crude oil demand will each increase by 32 percent (1.5 percent annually), the largest gain being in demand for natural gas.
- U.S. crude production will decline at an average annual rate of 0.2 percent causing imports to increase from 60 percent to 72 percent of consumption.
- Average crude prices will be $24.68 in 2000 dollars, an

average yearly increase of 0.6 percent; natural gas production will rise but prices (adjusted for inflation) will decline.

- Gross domestic product will grow by 3 percent and productivity by 2 percent annually, not necessarily robust, but significantly improved over the sluggish years of the 1970s and 1980s.

Of greater concern to Flying J are projections of motor fuel consumption. EIA expects gasoline consumption to increase by an average of 1.9 percent per year for the next 20 years and distillate by 3 percent. On the positive side for Flying J, 94 percent of the distillate increase is for diesel fuel, reflecting the anticipated growth in freight transportation. Various studies show that the trucking and diesel sales have a brighter future than most other elements of the petroleum industry.

Rather alarming for this country is that we consume one-quarter of the world's energy and hold only 2 to 3 percent of known world oil reserves compared to 66 percent for countries in the volatile Middle East. As national oil production continues to decline, we remain at the mercy of foreign governments for this energy source. This concern heightens when consideration is given to estimates that 65 percent of this country's known accessible oil supply has been consumed. These statistics are one reason why the federal government has recently accelerated programs to develop alternative power sources for motor vehicles. (Gasoline accounts for almost 45 percent of all petroleum used in this country.) Hybrid cars that marry a gasoline engine to an electric motor are currently on the highways as well as some vehicles powered by natural gas, but the most promising long-run development is fuel cell technology.

Fuel cells (first developed for NASA's space program) operate like batteries by using stored hydrogen and oxygen from the air to produce electricity. This technology gives promise of making cars and trucks cheaper to operate and pollution free. Through the Department of Energy, a partnership has been formed between the government and several auto producers in a program known as Freedom Cooperative Automotive Research. This alliance was established to accelerate the development of fuel cell applications as the power source for vehicles. In January of 2002, General Motors displayed an experimental model, and Ford and Chrysler hope to start limited production in 2004. However, the target year for making these models competitive with the internal combustion engine is 2020. One of the obstacles that could delay widespread usage is lack of a hydrogen refueling infrastructure, an issue Flying J is sure to address in the next few years.

Oil's share of the U.S. economy now accounts for 3 percent of GDP, down from almost 9 percent in the late 1970s. Although petroleum is growing less important to American industry as we move from manufacturing to a service and information economy, no one should underestimate the vital role it plays in greasing the gears of the world economy. Low oil prices reduce inflation and increase productivity while high prices restrict consumption and drain off purchasing power. And, as Federal Chairman Alan Greenspan warned in early 2002, "All economic downturns in the United States since 1973 have been preceded by sharp increases in the price of oil."

What does the future hold for Flying J? The straight-line projections noted above are certain to be disrupted by the inevitable volatility characteristic of the industry. One of Flying J's strengths is that it has learned to be amazingly comfortable during these troubled times. Most competitors have not held up as well, and many are facing a bleak future. Flying J will continue to grow at rates five times or so faster than the industry, placing some others in the death spiral of declining sales.

As noted, industry consolidation is inevitable. The trend in all service industries is for large firms, such as Wal-Mart and McDonald's, to drive out mom and pop stores and small chains too undersized to compete. Franchising, standardization, branding, and discounting have become the norm. The same degree of consolidation has not yet occurred within the travel plaza business, but the trend is definitely in that direction. As Phil stated, "I think there is going to be some consolidation. . . . If we talk in broad terms, this is still an industry undergoing a lot of change, still sorting itself out. And with change comes opportunity." These dynamic conditions can only enhance Flying J's position because of its size, management competence, and leadership position.

No other travel plaza company has come forth to seriously challenge Flying J, and none offers the same broad range of services. Flying J has the industry experts in travel plaza management and hospitality, and is fortunate to have some of the industry's most talented innovators in telecommunications and electronics. This is not to presume that the road ahead will be clear without its challenges for there is always the chance—given the company's current debt and the tendency to undertake new ventures each year—that it could become too overextended.

As Flying J nears the completion of constructing the last large format plazas, many ask, what will follow? Those who know Phil realize that every facet of travel hospitality is fair game. The company is looking forward to building several hundred smaller plazas along major highways that are not directly part of the interstate system. Bigger moves are more likely than mere brick and mortar additions. Like many ventures of the past, onlookers can expect to be taken off guard with new company developments. Some expect Flying J again to place all assets on the table in a gamble for a major takeover or in pursuit of a new venture. On several occasions, Phil has made known his desire to have Flying J's logo stand alongside those of the petroleum majors.

One approach Flying J will continue to rely on is forming partnerships with large corporations such as the current agreements with Conoco and Shell. Such practices are becoming increasingly common in the business world. Shell recently has been setting up dozens of business alliances with a wide range of partners. Shell's CEO Philip J. Carroll reasons that "the collaboration results in having better technology." Gaining capital is Flying J's primary concern. The quantity of available capital has determined the velocity of Flying J's advance throughout its history, a condition that continues to pace its advance. The company's need for financial partners is important in fully exploiting its current business lines including developing a chain of on-site motels, adding capitalization to the bank, purchasing gas and oil properties, and acquiring other companies that are being squeezed out in the inevitable consolidation occurring in every phase of the industry. Potential candidates include such diverse businesses as trucking, third-party billing companies, retail travel plazas, C-stores, refineries, insurance, telecommunications, and various aspects of food services.

Without question, the stunning growth of the past will not give way to complacency and conservatism. In considering the future, Phil likes to quote President Reagan, "Our best days are not behind us." Phil has been on record more than once in recent years with statements such as the following:

> I have never seen a time when we have had more opportunities to explore in all of our operating segments. Compared to what the future holds, you haven't seen anything yet!

FLYING J'S
CORPORATE PHILOSOPHY

We are in the hospitality and service business and our success is determined by: being able to not only meet but exceed our guest/customers' expectations; creating a work environment that attracts and holds capable, conscientious, honest and dedicated employees; and maintaining a relationship with our suppliers where they can grow and participate in our success.

Our Guests/Customers:

We aim to provide our guests/customers with friendlier and more courteous service than is available elsewhere. Furthermore, we will market high quality products delivered from facilities second to none. We will fairly and competitively price all products and services. We will treat our guests/customers as we would like to be treated.

Our Employees:

Our employees are the key to our success. To attract and hold quality people, we strive to pay competitive wages, offer attractive fringe benefits (by industry standards) and, most of all, provide a challenging, high integrity environment with unlimited growth opportunities where rewards are based on performance.

We require high standards of employee performance. We believe in rewarding people according to their abilities and the time and effort they put forth. We intend to share company profits with those who help produce them. Employees are to be judged on their performance in meeting company objectives. There is absolutely no place for poor, substandard performance, which in the long run hurts all employees as well as our guests/customers.

We expect employees to be honest and loyal to the company and for them to live by company rules and regulations. Employees should be guided by a sense of personal honor. If any employee has a question about a course of action, the answer should instinctively be: Do what is ethically right and morally proper. Dishonesty cannot and will not be tolerated at any level in our relationships with guests/customers, suppliers, other employees, or the company. We want positive, active thinkers because we recognize we live in a time of constant change. Therefore, we will strive to provide an atmosphere where suggestions for improvement are freely exchanged, evaluated, and acted upon. We must seize and exploit every new opportunity to get the job done in a more effective and efficient way. We will maintain an "open door" policy. If any employee thinks he or she has been treated unfairly, the employee can take his or her complaint to the next higher official—up to the company president.

Collectively, we must have an intense desire to be number one. We must all be builders. Decisions must be company decisions, not one person's decisions. We must put all the facts on the table, make the decision, and then have everyone support it. There is no room for backbiters and second guessers who do not get their own way.

In order to provide career opportunities for our people, we must commit ourselves to a sound, planned pattern of growth. To finance and grow, we must have profit. And, to be profitable, we must have loyal and productive employees who want to succeed and want the company to succeed.

Company management makes this commitment to all employees: No employee will ever be treated as a number and each employee will be treated with dignity and respect.

Management realizes that growth creates extra demands on the time of key individuals creating situations that sometimes can be misinterpreted as lack of interest in the welfare of an individual or group of employees. It is during these times that members of the management team must rededicate themselves to the philosophy that all employees will treat other employees, guests/customers, and suppliers as they would like to be treated.

Our Suppliers:

Our suppliers are key to our long-term success. We must be fair and honest with them. Remember that in order for a relationship to last and prosper, it must fill a need of each party involved. We expect top performance with our suppliers and commit ourselves to perform at the same level.

Our Company:

Our company must deserve and earn the respect of our guests/customers, suppliers, competition and, most of all, our employees. We want a company where everyone is proud to say "I helped make Flying J what it is!"

INDEX